普通高等教育“十三五”规划教材

# 有色冶金科技英语写作

谢 锋　王 伟　蒋开喜　编著

北 京
冶 金 工 业 出 版 社
2020

## 内 容 提 要

本书以东北大学有色冶金专业方向研究生教学体系为基础，结合有色冶金专业基本概念和工艺特点，详细介绍了有色冶金科技英语写作中词汇和词组的用法以及句式表达技巧；结合本专业科技论文的写作惯例，详细介绍了专业文献检索工具与使用方法、英文论文的格式编辑以及投稿方法等；结合本专业代表性论文的撰写和修改案例，演示了强化和提高高水平英文期刊论文写作能力的知识和技巧。

本书具有较强的专业性和实用性，可作为高校冶金工程相关专业教学用书，也可供相关专业研究学者和工程技术人员参考。

**图书在版编目(CIP)数据**

有色冶金科技英语写作/谢锋，王伟，蒋开喜编著.—北京：冶金工业出版社，2020.11

普通高等教育“十三五”规划教材

ISBN 978-7-5024-8639-6

Ⅰ.①有…　Ⅱ.①谢…　②王…　③蒋…　Ⅲ.①有色金属冶金—英语—写作—高等学校—教材　Ⅳ.①TF8

中国版本图书馆 CIP 数据核字(2020)第 216724 号

出 版 人　苏长永

地　　址　北京市东城区嵩祝院北巷 39 号　邮编　100009　电话　(010)64027926

网　　址　www.cnmip.com.cn　电子信箱　yjcbs@cnmip.com.cn

责任编辑　张熙莹　郭雅欣　美术编辑　彭子赫　版式设计　禹　蕊

责任校对　李　娜　责任印制　李玉山

ISBN 978-7-5024-8639-6

冶金工业出版社出版发行；各地新华书店经销；三河市双峰印刷装订有限公司印刷

2020 年 11 月第 1 版，2020 年 11 月第 1 次印刷

787mm×1092mm　1/16；11.5 印张；276 千字；174 页

**38.00** 元

**冶金工业出版社　投稿电话　(010)64027932　投稿信箱　tougao@cnmip.com.cn**

**冶金工业出版社营销中心　电话　(010)64044283　传真　(010)64027893**

**冶金工业出版社天猫旗舰店　yjgycbs.tmall.com**

（本书如有印装质量问题，本社营销中心负责退换）

# 前　言

有色冶金工业作为重要基础产业之一，在国民经济发展和国家安全保障等方面发挥着极其重要的作用。我国在有色冶金领域的基础研究和技术应用已趋于国际领先水平。在建设一流学科和一流高校的时代背景下，我国高等教育学科建设与教材建设的改革不断深入，有色冶金学科高级人才的培养面临着国际化、信息化和专业化等多种需求的挑战。英语在科学研究以及国际交流中的作用日益显著。为了进一步深化有色冶金专业方向教学改革，优化课程体系，推进有色冶金专业研究生课程的教学实施，结合多年的教学经验，作者编写了这本具有较强专业特色的有色冶金科技英语写作教材。

有色冶金学科研究范围广泛，很多研究方向都有较强的专业特色。本书以东北大学有色冶金专业方向教学体系为基础，结合有色冶金专业基本概念和工艺特点，力图从词汇到常用表达句式以及写作格式等方面突出专业特色。本教材编写的过程中，作者参阅并引用了国内外有关文献，吸收和借鉴了一些教材的精华，在书后一一列出，在此对文献作者表示衷心的感谢。加拿大 UBC 大学 David Dreisinger 教授对本书部分内容的修订给予了大力支持，此外，东北大学张壮博士和白云龙博士为本书的校对做了大量工作，作者在此深表感谢。

本教材主要供冶金工程专业本科生和研究生教学使用，也可供相关专业教学、工程技术人员培训及相关专业研究和设计人员参考使用。

由于水平所限，书中存在的不足之处，恳请批评指正。

作　者

2020 年 5 月

# 目　录

# 1 绪　论

## 1.1 科技英语写作的特点

有色冶金工业作为重要基础产业之一，在我国国民经济发展和国家安全保障等方面发挥着极其重要的作用。随着有色冶金行业的不断发展壮大，我国有色冶金领域的基础研究和技术应用已趋于国际领先水平。在本领域的国际交流也变得愈发重要。有色冶金的从业人员不但需要持续吸收国际先进科技成果和信息，也需要向世界介绍和传播自己的研究成果。因此，科技英语的写作能力的培养日趋重要。与一般的基础英文写作相比，科技英语写作具有较鲜明的专业特点，主要体现在词汇、时态、语态及文章结构等方面。

### 1.1.1 时态

从时态的角度看，由于科技英语写作侧重于叙述科技内容，很少涉及动作发生的时间，因此科技写作中时态运用很有限，常用的只有几种。一般现在时是最常用的时态，其次是一般将来时和现在完成时，再次是一般过去时和现在进行时。科技英语文章中不同部分选用的时态也有所不同，如在引言中用一般现在时描述一些客观真理或事实等。如：

The fundamental principles of the extraction behavior of gold are discussed.

本文讨论了金提取行为的基本原理。

An empirical formula for copper leaching is also given.

同时给出了铜浸出的经验公式。

Several simple dielectric test objects have been utilized.

我们使用了几种简单的介质实验目标。

This scheme avoids the complexities of matrix inversion.

本方法避免了矩阵求逆的复杂运算。

The results obtained demonstrate that the above equation holds for all cases.

所得结果表明，上述公式适用于各种情况。

### 1.1.2 语态

从语态的角度看，科技英语写作在描述一些行为时常选择被动语态，这是因为被动语态能够省略主语，增强文章的客观性。在科技论文中由于着重讲述客观现象和科技真理，因此与日常英语和文学著作相比被动句使用得广泛得多，而在英文文摘中更是如此。汉语中被动句用得较少，且可使用没有行为主体的无人称句，如果用英语表达时就应使用被动句，否则很容易写出没有主语的“中式英语”。当然科技英语中也有表示“人们、大家”的词，即“one”。这里需要注意的是，“people”在正式的科技英语写作中是很少用的。

下面举一些例子。

Nano materials based on silver have been widely used in our national defense.

银基纳米材料已广泛地用于我们的国防。

Finally, the application of this recovery method is illustrated with an example.

最后举例说明了这种回收方法的应用。

It is well known that molecules consist of atoms.

众所周知，分子是由原子组成的。

A new computer-aided design method is presented.

本文提出了一种新的计算机辅助设计方法。

需要注意的是，尽管副词作状语时在句中的位置是灵活的，但在被动句中修饰主动词的副词大多应放在过去分词之前，而不放在句首，如：

The leaching circuit for this copper ore is then completed.

针对这种铜矿石的浸出流程于是就完成了。

Wires are usually made of copper.

导线通常是由铜制成的。

The current in this anode is first determined.

首先求出这个阳极的电流。

### 1.1.3 句子特征

科技英语文章中的句子特征有：

（1）句子较长，句子的结构严密紧凑。中文句子一般比较简短，看起来似乎比较松散；而科技英语文章的句子与日常用语、文学语言相比，结构严密且比较冗长，从而节省篇幅。流畅优雅的长句对于习惯阅读、研究科技内容的科技工作者来说是合适的，因为科技英语文章要求使用完整和严密的句法，句子成分多、联合成分多、层次多、复句多、长句多，经常出现多个分句并列或复合句中从句套从句的现象。同时还频繁使用祈使句，“it”句型、动词的非谓语形式等，从而达到表述上的清楚、简洁、准确。若使用过多的简单句，会使逻辑关系不紧密，文章读起来单调乏味。但句子不能过长，否则会使读者阅读困难。总之论文中的句子结构应该多样化，长短句兼有。当然，在科技文中的句子的长短还取决于文章的种类。文献的学术性越强，概念越深奥的问题，则使用的句子就越长。有统计资料表明，科技写作的句子平均长度在20~30个词之间。

（2）并列复合句中往往文体应一致。文体要一致，就是指并列复合句中各分句一般来说应使用同一时态、同一语态，有时甚至应使用同一句型；特别是不要前一句分句用主动句，而后一分句用被动句。

（3）静态结构。不少情况下往往采用静态结构（stative construction）来表示状态、特征、情况，以取代采用行为或过程的句法结构。常见的有以下两种句型：

1）连系动词+表语：

This pH electrode is breakable. (=can be broken)

这个pH电极容易打碎。

This concept is applicable to metallic conductors. ( = can be applied)

这个概念可应用于金属导体。

This exponential function is absolutely integrable. ( = can be integrated)

该指数函数是绝对可积的。

This phenomenon is ignorable. ( = can be ignored)

这个现象可以忽略。

The relationship between aluminum extraction and leaching time is expressible as follows. ( = can be ignored)

铝的提取率和浸出时间可表示如下。

2）趋于使用某些静态动词：

The high voltage leads to( = causes) the breakdown of the anode.

高压导致了阳极毁坏。

Nearly all substances fall into( = are divided into) two categories.

几乎可以把所有的物质分成两类。

（4）用短语代替从句。不少情况下用短语代替从句（特别是代替状语从句和定语从句），可防止句子显得松散无力，成为使句子结构紧凑的一种手段。

1）用介词短语代替状语从句：

原：... because it interacts with the ligand.

改：... because of its interaction with the ligand.

……因为它与络合剂相互作用的缘故。

原：... because the scattering cross section of the target is large.

改：... because of the large scattering cross section of the target.

……由于目标的散射横截面积比较大。

原：After they had substituted the actual magnitudes, A turned out to be the velocity of light.

改：Upon substitution of the actual magnitudes, A turned out to be the velocity of light.

把实际的数值带入后，发现 A 是光速。

2）用作主语的名词短语代替表示条件、原因、时间、目的等的状语从句：

原：If the cathode surface is analyzed physically it can be shown that...

改：A physical analysis of the cathode surface shows that...

如果对阴极表面进行物理分析的话，可以看出……

原：If Q is large, a very sharp resonant curve can be obtained.

改：A large Q leads to a very sharp resonant curve.

如果 Q 值较大，就可以获得很锐的谐振曲线。

原：If KVL is applied to the collector-emitter loop, the following simple equation can be obtained.

改：Application of KVL to the collector-emitter loop yield/produces/gives/leads to/gives rise to/results in the following simple equation.

如果把基氏电压定理（KVL）应用于集电极-发射极回路，就可以得到下面的简单方程式。

原：If the assumptions are thoroughly examined, it can be shown that…

改：A thorough examination of the assumptions reveals that…

如果透彻地考察一下这些假设，就能揭示出……

原：Since there is no atmosphere in space, scientists could make pure species there.

改：The absence of atmosphere in space would enable scientists to make pure species there.

由于太空中没有大气，因此科学家们能够在那里制造纯净物质。

原：Because the iron is present in the coil, the magnetic induction has been increased to over 5500 times what it was.

改：The presence of the iron in the coin has increased the magnetic induction to over 5500 times what it was.

由于在线圈中有铁存在，使磁感应提高到原来的5500多倍。

原：With the advent of large scale integration, it has been possible to create economically tens of thousands of logical, arithmetic, and storage devices on a tiny semiconductor chip.

改：The advent of large scale integration has made it possible to create economically tens of thousands of logical, arithmetic, and storage devices on a tiny semiconductor chip.

由于大规模集成的出现，人们能够在一块微小的半导体芯片上经济地制造出成千上万个逻辑器、运算器和储存器。

原：After the chloride is added, the output of copper extraction becomes stable.

改：The addition of the chloride makes the output of copper extraction stable.

添加氯化物后，铜的提取量稳定了。

原：In order that the electrical operation of the ferrites can be successful, it is demanded that…

改：Successful electrical operation of the ferrites demands that…

为了使铁氧体的电气作用能够成功，就要求……

3）其他情况。在科技英语写作中，用分词短语、分词独立结构、“with”短语及“with”结构作状语比较普遍，放在句首通常可用来代替时间、条件、原因等状语从句使句子结构紧凑；放在句末主要用来避免使用“and”罗列并立的事情，这在叙述事物如何进展时特别有用，可防止句子显得松散无力。如：

Using a SEM (=If we use a SEM), we can observe objects far small.

如果使用SEM，我们就能观察很小的东西。

With no resistance in the cell circuit(=If there were no resistance in the cell circuit), the current would increase forever.

如果电解槽电路中没有电阻，电流就会无限增大。

In these graphs the horizontal axis measures time, increasing toward the right away from the vertical axis.

在这些曲线图中，水平轴度量的是时间，它朝右边移离垂直轴是增加的。

In this case copper ions will possess more energy, thus increasing the oxidation of gold.

在这种情况下，铜离子将具有更多的能量，因而增加了金的氧化。

用介词短语、形容词短语、分词短语等作后置定语或用现在分词或动名词等作前置定语代表定语从句等，使句子精简。

There are many species larger than the copper ion. ( =that are larger than the copper ion)

有许多物质比铜离子大。

The building under construction ( =that is being constructed) will be a material research institute.

正在建造的那栋大楼将是材料研究所。

A capacitor is a device consisting of ( =that consists of) two conductors separated by( =that are separated by) an insulator.

电容器是由被绝缘体隔开的两个导体组成的一种器件。

It is necessary to determine the current flowing in this cell circuit. ( =which flows in this circuit)

必须求出在这个电解槽电路中流动的电流。

Those are the rays reaching the counter face. ( =which reach the counter face)

那些是到达背面的光线。

科技英文中经常出现名词短语作前置定语的情况，因为它具有简洁性。一般来说，论题的技术性越强、越专门化，这种复合名词出现得就越频繁、越复杂，特别在科技杂志、科技新闻标题和技术广告中尤为如此。如：

the University of Utah artificial arm

犹他大学的人造臂

the Ministry of Education Key Laboratory

教育部重点实验室

an order of magnitude improvement

提高了一个数量级

a signed decimal 10’s complement converter

带符号十进制10的补的转换器

a diesel engine transmission unit

内燃机变速装置

（5）常用的说法。有一些固定的常用的表达方法，可以使英语写作表达得更地道。以下仅举几例。

1）The leaching device is small in size.

这种浸出设备的体积小。

从语法上讲，上句也可译为“The size of this leachin device is small.”但英美科技人员偏爱上述句型；另外这里的“体积”有不少读者将其译为“volume”，这样翻译不够准确，因为该词强调“容积”之意，以及用来表示固体、液体或气体所占的空间等。

2）Emphasis is put on what is described in this chapter.

重点放在这一章所讲的内容上。

由于在“what 从句”中用了被动语态，这时只能用“in this chapter”而不能用“by this chapter”，尽管使用主动语态时可以说“what this chapter describes”，这是英美人的习惯。另外，这里的“内容”不要写成“content”，要不然就显得不地道，这也是一些读者喜欢用的词汇。

3）This leaching device is characterized by its small size, light weight, high efficiency and low cost.

The leaching device is small in size, light in weight, high in efficiency and low in cost.

The features of this leaching device are small size, light weight, high efficiency and low cost.

这种浸出设备的特点是体积小、质量轻、效率高、成本低。

这句话可以有以上三种表达形式，其中第一种最佳。但不可以表示成：The features of this leaching device are small in size, light in weight and low in cost.

4）Current and voltage are related by the following equation.

或 Current is related to voltage by the following equation.

电流与电压之间的关系可以用下式表示。

该句型中的“A and B are related by C”或“A is related to B by C”来表示“A 与 B 之间的关系由 C 表示”。又如：

An increase in temperature is related to the copper consumption by the following equation.

温度的上升与铜消耗之间的关系由下式表示。

The maximum junction temperature and maximum dissipation are related by the curve shown in Fig. 1-2.

最高结温与最大功耗之间的关系由图 1-2 所示的曲线表示。

The dissolution of nickel is related to the activity gains by Debye-Hucel equation.

镍的溶解与德拜-胡克公式得到的活度有关。

5）在表示“……的方法是……”，英美科技工作者往往不是采用“The method for…is…”而是常采用“…is found/minimized/determined/increased/solved by…”这一句型。如：

This distortion is minimized by proper leaching circuit design and suitable choice of the operating conditions.

使这种误差降至最小的方法是通过合适的浸出流程设计和选择适当的操作条件。

The operating point is determined by writing the equation for the load line.

确定工作点的方法是写出负载线的方程式。

The average power is found by averaging Eq. (2-6) over a complete cycle.

求平均功率的方法是对式（2-6）取完整一周内的平均值。

The capacitance is creased by making the area of the plates large.

增大电容量的方法是加大平板的面积。

The current in the circuit is obtained by first computing the total complex impedance.

获得这一条件的方法是把一个高阻抗负载连接到输出端。

This condition is satisfied by connecting a high-impedance load to the output terminals of the electrolytic cell.

满足这一条件的方法是把一个高阻抗负载连接到电解槽输出端。

Copper loss is reduced by increasing the value of the filter capacitor.

减小铜损失的方法是增大过滤器容积。

（6）通常避免口语体的缩略形式。科技英语文章属于正规的书面文体，所以在其中一般不用口语体的缩略形式。

原：It's of great importance to...

改：It is of great importance to...

原：It's been shown that...

改：It has been shown that...

原：Next, we'll deal with...

改：Next, we shall deal with...

原：We can't determine...

改：We cannot determine...

此外，科技英语写作的文章常有固定的结构。不同期刊对文章的结构可能略有差异，但总体都是大同小异。通常都会有摘要、引言、实验方法、实验结果与讨论和结论几个部分。每部分在本书后续章节中均有介绍。对于科研工作者来说，为了将自己的研究成果和经验在国际上进行交流，必须要掌握这些技巧。

## 1.2 英语科技论文结构

通常而言，英语科技论文应当做到概念明确、判断恰当、推理合乎逻辑，同时应保持科学的严密性。不同学科的科技英语论文的结构有所差别。对于有色金属冶金相关期刊而言，文章主体部分大多可以分为摘要（Abstract）、引言（Introduction）、实验方法（Methodology）、结果与讨论（Results and Discussion）和结论（Conclusion）几个部分。除了文章主体部分外，还有文章标题（title）、作者的姓名（author's name）、作者的工作单位（institution）、致谢部分、参考文献、作者简介等部分。本小节将依次介绍各部分的写作要求与写作原则。

### 1.2.1 标题

标题（title）应写得简明扼要，同时能够吸引读者。写标题时，切勿过于笼统，此外字数不宜过多。

论文的标题应该使用名词短语，不要使用一个句子或不定式短语；一般也不用介词短语等形式（除了“on 短语”外），同时在标题中不能出现从句。如：

Parametric Up-Convertion of Gold-based Nano-material

Comparative Study of Kinetic Control by Iron thiosulfate Leaching System

Electron Trigonometry—a New Tool for Electron-Optical Design

The Effect of Nonsymmetrical Doping on Tunnel Diodes

An Extended Definition of Catalytic Leaching

在少数情况下，发现有用动名词短语，或用“on”引出的介词短语，表示“论（关于）……”之意。如：

Improving the Stability of Copper-ammonia Leaching System

On the Cascaded Tunnel-Diode Amplifier

杂志正式发表论文时，编辑会在标题的右上方打一个＊号，然后在该页的脚注中注明论文收到的时间及经重新修改后寄达的时间。如：

Manuscript received February 9，2009；revised August 1，2009.

Received by the IRE，April 21，2010；revised manuscript received，October 28，2010.

另外，标题中开头的冠词可以省去（我国的期刊论文写作要求省去标题中开头的冠词。在写作者的通信地址时，其组织机构名称前的定冠词也常省去；英美国家大多数科技杂志名称前一般不加定冠词）。关于标题的大小字母表示法，通常有以下三种形式：

（1）开头第一个字母大写、专有名词大写，其余均采用小写字母，如：

A novel leaching system for copper extraction from chalcopyrite

（2）开头的字母和每个实词及不少于5个字母的介词、连词的第一个字母均大写，如：

Study of a Gold Leaching System Formed by Ammonia and Thiosulfate

（3）全部字母均大写（这种形式一般用于计算机检索系统），如：

A DISCUSSION ON SELF-ADAPTIVE SYSTEMS

### 1.2.2 作者的姓名

作者的姓名（author's name）一定要用全称，这主要是针对外国人而言，英文他们的姓名一般由三部分组成，即名（first name），中间名（middle name），姓（last name）。也有两部分或四部分构成的。

对于中国人的姓名，不少人也按外国人的姓名顺序写法表示。如“王伟”可以写成W. Wang，Wei Wang。有些著名的期刊（如IEEE-Institute of Electrical and Electronics Engineers）要求在作者姓名后需要加会员名称（如果有）。如：

J. C. HELMER, STUDENT MEMBER, IEEE

（IEEE 实习会员）

AARON A. GALVIN, MEMBER, IEEE

（IEEE 会员）

LEONARD R. KAHN, SENIOR MEMBER, IEEE

（IEEE 高级会员）

R. L. McFARLAN, FELLOW, IEEE

（IEEE 会士）

### 1.2.3 作者的工作单位

有些杂志把工作单位（institution）直接标注在作者姓名下面，如：

Department of Information Engineering, Xidian University, Xi'an 710071, China

Cambridge University Press, 40 West 20$^{th}$ Street, New York, NY 10011-4211, USA

需要注意的是，我们的邮编写在城市或省名之后，其间不用逗号，只要空一格即可。也有不少杂志把工作单位写在论文第1页的脚注中，如：

The author is with（部分人用“at”）the Department of Applied Mathematics, School of

Science at Xidian University, Xi'an 710071, China.

The authors are with the Department of Electrical Engineering and Computer Sciences, and the Electronics Research Laboratory, University of California, Berkeley, CA 94720.

The author is with the Department of Electrical Engineering, University of Toronto, Toronto, Ontario, Canada M5S 1A4.

### 1.2.4 论文目录

有的期刊要求有论文目录，但更多的期刊是不要求列出这一项的。一般来说，比较长的论文应该附有内容目录，该目录就是论文正文部分的小标题并附有页码，这样可使读者一目了然地看出论文内容的大致轮廓，同时便于读者查阅某一内容。

### 1.2.5 文章主体部分

#### 1.2.5.1 摘要（Abstract）

摘要的主要功能是让读者快速地了解文章的研究意义、研究方法及研究结论。如果读者感兴趣，可以继续阅读全文，了解其中的细节。这一部分完全依靠文章的内容决定，因此可以认为摘要代表了全文。在实际写作时摘要一般最后再写。

对于作者而言，摘要写得是否精彩能够部分地决定文章是否吸引读者。一篇好的摘要应该能充分的反应文章的关键信息，并且能够引起读者的兴趣。摘要的结构和全文的结构类似，通常可以分为以下四个部分。

首先可以在摘要中描述文章的背景信息、研究目的等。其内容与引言部分类似，但是要更加精简。以下列举一些这部分常用的词汇。

| | | | |
|---|---|---|---|
| a number of studies | it is known that | exist(s) | it is widely accepted that |
| frequently | occur(s) | generally | often |
| is a common technique | popular | is/are assumed to | produce(s) |
| is/are based on | recent research | is/are determined by | recent studies |
| is/are influenced by | recently | is/are related to | recently-developed |
| it has recently been shown that | to examine | in order to | to investigate |
| our approach | to study | the aim of this study | with the aim of |
| to compare | | | |

其次是文章的实验方法和实验材料等。在摘要中进行概括性的描述即可。以下列举一些这部分常用的词汇和示例。

| | | | |
|---|---|---|---|
| was/were assembled | was/were calculated | was/were constructed | was/were evaluated |
| was/were formulated | was/were modeled | was/were performed | was/were recorded |
| was/were studied | was/were treated | | |

然后是实验的结果、主要结论或意义等。这部分的内容根据实验与讨论部分即可。以下列举一些这部分常用的词汇和示例。

| | | | |
|---|---|---|---|
| caused | was/were achieved | decreased | was/were found |

| | | | |
|---|---|---|---|
| had no effect | was/were identical | increased | was/were observed |
| it was noted/observed that… | was/were obtained | The evidence/These results… | it is thought that |
| indicate(s) that | we conclude that | mean(s) that | we suggest that |
| suggest(s) that | achieve | accurate | allow |
| better | demonstrate | consistent | ensure |
| effective | guarantee | enhanced | obtain |
| exact | validate | improved | compare well with |
| new | for the first time | novel | in good agreement |
| significant | The evidence/These results… | simple | indicate(s) that |
| suitable | mean(s) that | superior | suggest(s) that |
| applicability | make it possible to | can be applied | potential use |
| can be used | relevant for/in | | |

最后是实验的局限或未来展望等。以下列举一些这部分常用的词汇和示例。

| | | | |
|---|---|---|---|
| a preliminary attempt | future directions | slightly | not significant |
| future work | | | |

1.2.5.2 引言（Introduction）

写引言时你需要知道自己研究的是什么，有什么目的，自己做了什么工作等，因此引言部分通常需要在报告部分完成后再准备。引言可以分为四个部分。第一部分的内容通常包括研究方向的重要性，背景信息，解释关键词或文章中专有名词和目前该研究领域出现的问题或研究的方向等。首先谈谈研究方向的重要性，构建研究内容的重要性十分重要。描述研究方向的重要性常用的词汇和语句如下。

| | | | |
|---|---|---|---|
| (a) basic issue | economically important | (a) popular method | over the past ten years |
| (a) central problem | (has) focused(on) | (a) powerful tool/method | play a key role(in) |
| (a) challenging area | for a number of years | (a) profitable technology | play a major part(in) |
| (a) classic feature | for many years | (a) range(of) | possible benefits |
| (a) common issue | frequent(ly) | (a) rapid rise | potential applications |
| (a) considerable number | generally | (a) remarkable variety | recent decades |
| (a) crucial issue | (has been) extensively studied | (a) significant increase | recent(ly) |
| (a) current problem | importance/important | (a) striking feature | today |
| (a) dramatic increase | many | (a) useful method | traditional(ly) |
| (an) essential element | most | (a) vital aspect | typical(ly) |
| (a) fundamental issue | much study in recent years | (a) worthwhile study | usually |
| (a) growth in popularity | nowadays | (an) advantage | well-documented |
| (an) increasing number | numerous investigations | attracted much attention | well-known |
| (an) interesting field | of great concern | benefit/beneficial | widely recognized |
| (a) key technique | of growing interest | commercial interest | widespread |
| (a) leading cause(of) | often | during the past two decades | worthwhile |
| (a) major issue | one of the best-known | | |

Copper extraction from low grade ore have received much attention in recently years.

Much study in recent years has focused on...

A major current focus in oxidative leaching is how to ensure sustainability of...

Numerous experiments have established that ionizing radiation causes...

Low microwave radiation has generated considerable recent research interest.

Analysis of change in the diffusion sector is vital for two important reasons: ...

It is generally accepted that joints in steel frames operate in a semi-rigid fashion.

Convection heat transfer phenomena play an important role in the development of...

For more than 100 years researchers have been observing the stress-strain behavior of...

Much research in recent years has focused on catalytic leaching of refractory gold ore.

很多初学者没有信心认为自己的研究十分重要，但大多数学者在写论文时，首先会构建自己研究的意义，以强调研究的重要性。意义本质上是人对事物赋予的含义，而事物本身是没有意义的。所以，写自己的文章时不必吝惜，努力发掘研究领域和方向的意义及重要性吧。需要注意的是在描述意义时，通常使用现在完成时，这是因为现在完成时具有从过去持续到现在的潜在含义。

构建研究内容的意义后，可以描述研究的背景信息。描述背景信息通常使用一般现在时。这是因为一般现在时常用于描述客观现实，而背景信息包含的正是这样的内容。在描述背景信息之前，首先需要考虑论文的受众是谁，他们对文章研究的背景信息了解有多少？简言之，引言部分的背景信息需要考虑读者对背景信息的了解深度，以便读者阅读时能够迅速地掌握研究的背景。那么如何确定读者的了解深度呢？这和论文的研究方向有关。举例来讲，如果研究方向本身已经是非常细分的领域，那么可以从非常专业的背景信息开始。如果论文希望吸引到更大范围的读者，那么应该从一般化的背景信息开始。

除了研究的重要性和背景信息外，这一部分通常还需要描述一下该研究领域目前的研究方向或研究问题，以此引出引言的第二部分：过去或目前的研究进展。这个部分首先需要作者阅读大量文献，以至于对该领域的状况了如指掌。回顾文献时，不要只是把他人的研究成果陈列上去。作者需要提供一个逻辑框架陈述这些研究成果。常用的逻辑框架有时间顺序、不同的研究手段或不同的理论等。这样做的好处是便于读者阅读时能够快速掌握，也可以使文章读起来更有条理。以下列举一些回顾文献常用的词汇和示例。

| | | | |
|---|---|---|---|
| achieve | deal with | incorporate | provide |
| address | debate | indicate | publish |
| adopt | define | interpret | put forward |
| analyse | demonstrate | introduce | realise |
| apply | describe | investigate | recognise |
| argue | design | illustrate | recommend |
| assume | detect | implement | record |
| attempt | determine | imply | report |
| calculate | discover | improve | reveal |
| categorise | discuss | measure | revise |
| carry out | enhance | model | review |
| choose | establish | monitor | show |

claim
classify
collect
compare
concentrate(on)
conclude
conduct
confirm
consider
construct
correlate
estimate
evaluate
examine
explain
explore
extend
find
focus on
formulate
generate
identify
note
observe
prefer
overcome
perform
point out
predict
present
produce
propose
prove
simulate
solve
state
study
support
suggest
test
undertake
use
utilise

This phenomenon was demonstrated by…

In their study, expanded T-cells were found in…

Initial attempts focused on identifying the cause of…

Weather severity has been shown to…

Early data was interpreted in the study by…

The algorithm has been proposed for these applications…

The results on pair dispersion were reported in…

Their study suggested a possible cause for…

An alternative approach was developed by…

引言的第三部分是寻找过去的研究与目前研究中的不足或问题，而你的研究正是要解决这些问题。过去的研究由于研究手段和研究方法受限，可能多少有些不足。而你的工作，就是解决这些问题。以下列举一些这部分常用的词汇和示例。

ambiguous
confused
doubtful
false
ill-defined
improbable
inadequate
incompatible(with)
inconclusive
inconvenient
inefficient
inflexible
meaningless
non-existent
not apparent
not repeatable
not sufficiently+adjective
(the) absence of
(a) challenge
(a) difficulty
(a) drawback
(a) flaw
(a) lack
(a) need for clarification
no correlation(between)
(a) problem
(a) weakness
(to) demand clarification
(to) fail to
(to) miscalculate
(to) misunderstand
(to) neglect
(to) remain unstudied
(to) suffer(from)
computationally demanding
deficient
expensive
far from perfect
impractical
inaccurate
incapable(of)
incomplete
inconsistent
incorrect
inferior
insufficient
misleading
not addressed
not dealt with
not studied
not well understood
(an) alternative approach
(a) defect
(a) disadvantage
(an) error
(a) gap in our knowledge
(a) limitation
(the) next step
(an) obstacle
(a) risk
(to be) confined to
(to) disagree
(to) fall short of
(to) misjudge
(to) need to re-examine
(to) overlook
(to) require clarification
few studies have…

| | | | |
|---|---|---|---|
| not/no longer useful | it is necessary to... | of little value | little evidence is available |
| over-simplistic | little work has been done | poor | more work is needed |
| problematic | there is growing concern | questionable | there is an urgent need... |
| redundant | this is not the case | restricted | unfortunately |
| time-consuming | unproven | unanswered | unrealistic |
| uncertain | unresolved | unclear | unsatisfactory |
| uneconomic | unsolved | unfounded | unsuccessful |
| unlikely | unsupported | unnecessary | |

Few researchers have addressed the problem of...

There remains a need for an efficient method that can...

However, light scattering techniques have been largely unsuccessful to date.

Unfortunately, these methods do not always guarantee...

The high absorbance makes this an impractical option in cases where...

An alternative approach is necessary.

The function of these lixiviants remains unclear.

These can be time-consuming and are often technically difficult to perform.

在这个部分中还需说明文章的研究目的。在描述研究目的时，通常使用一般过去时，如：

The aim of this project was...

在引言的最后一部分，你需要介绍你的研究工作，也就是这篇文章内容的概括。其中可以包括论文的结构、研究方法、主要发现等。需要注意的是，这里不必写得过于详细，因为你的实验方法、实验结果主要写在后续部分。以下列举一些描述工作内容常用的词汇和示例。

| | | | |
|---|---|---|---|
| (to) facilitate | (this) work | (to) illustrate | begin by/with |
| (to) improve | close attention is paid to | (to) manage to | here |
| (to) minimise | overview | (to) offer | simple |
| (to) outline | straightforward | (to) predict | successful |
| (to) present | valuable | (to) propose | aim |
| (to) provide | goal | (to) reveal | intention |
| (to) succeed | objective | | |

This paper focuses on...

The purpose of this study is to describe and examine...

In order to investigate the biological significance...

In this paper we present...

New correlations were developed with excellent results...

In the present study we performed...

This paper introduces a scheme which solves these problems.

The approach we have used in this study aims to...

#### 1.2.5.3 实验方法（Methodology）

这部分在不同的期刊中可能有不同的叫法，如 Materials & Methods, Procedure, Experiments, Experimental, Simulation and Methodology or Model 等。具体的要求可以在期刊官方网站的 Guide for Authors 中找到。一般期刊对实验方法这一节的要求为：为读者提供足够的细节，以至于可以重复实验并得到相似的结果。实验方法通常可以拆解为三个部分，第一个部分可以对实验材料或实验方法进行概括性描述，让读者知道接下来会介绍什么。除此以外，还应该说明实验所使用的设备仪器等。之所以在介绍实验方法前进行概括性介绍，是为了使读者阅读时更容易概括性地了解这部分内容。以下列举一些概括性描述实验方法常用的词汇和示例。

| | | | |
|---|---|---|---|
| all(of) | most(of) | both(of) | the majority(of) |
| each(of) | (the) equipment | many(of) | (the) chemicals |
| (the) tests | (the) models | (the) samples | (the) instruments |
| (the) trials | (the) materials | (the) experiments | is/are commercially available |
| was/were acquired(from/by) | was/were devised | was/were carried out | was/were found in |
| was/were chosen | was/were generated(by) | was/were conducted | was/were modified |
| was/were collected | was/were obtained(from/by) | was/were performed(by/in) | was/were supplied(by) |
| was/were provided(by) | was/were used as supplied | was/were purchased(from) | was/were investigated |
| opposite | facing | out of range(of) | within range(of) |
| below | under | above | over |
| parallel(to/with) | perpendicular(to) | on the right/left | to the right/left |
| (to) bisect | (to) converge | near side/end | far side/end |
| side | edge | downstream(of) | upstream(of) |
| boundary | margin | on the front/back | at the front/back |
| higher/lower | upper/lower | horizontal | vertical |
| circular | rectangular | equidistant | equally spaced |
| on either side | on both sides | is placed | is situated |
| is mounted(on) | is coupled(onto) | is aligned(with) | is connected(to) |
| extends | is surrounded(by) | is attached to | is covered with/by |
| underneath | on each side | on top(of) | is located |
| adjacent(to) | is fastened(to) | border | is fixed(to) |
| in the front/back | is fitted(with) | inner/outer | is joined(to) |
| lateral | occupies | conical | is positioned |

The impact tests used in this work were a modified version of…

All reactions were performed in a 27mL glass reactor…

The base lines were generated as previously described in…

In the majority of the tests, buffers with a pH of 8 were used in order to…

The cylindrical lens was obtained from Newport USA and is shown in Fig. 3.

The experiments were performed in a sealed cell so that…

The substrate was obtained from the MATACT Research Centre…

A high-resolution telescope were used in this study to perform...

Porosity was measured at the near end and at the far end of the polished surface.

The compression axis is aligned with the rolling direction...

The source light was polarized horizontally and the sample beam can be scanned laterally.

The intercooler was mounted on top of the engine...

The mirrors are positioned near the focal plane.

The intercooler was mounted on top of the engine...

The concentration of barium decreases towards the edge...

在概括性地描述过实验方法后，就可以详细介绍实验方法的细节了，如实验的温度、实验流程或实验条件等。在这一部分中，除了实验细节外，还应该包括一些实验方法的背景信息，比如选择这种实验方法的原因，某一步骤为什么要这么做等。这是因为一些对你显而易见的原因，对于读者而言可能不是那么容易察觉。你要让你的读者接受你实验方法的合理性，以此增加文章的说服力。以下列举一些这部分常用的词汇和示例。

| | | | |
|---|---|---|---|
| was adapted | was divided | was added | was eliminated |
| was adopted | was employed | was adjusted | was estimated |
| was applied | was exposed | was arranged | was extracted |
| was assembled | was filtered | was assumed | was formulated |
| was attached | was generated | was calculated | was immersed |
| was calibrated | was inhibited | was carried out | was incorporated |
| was characterised | was included | was collected | was inserted |
| was combined | was installed | was computed | was inverted |
| was consolidated | was isolated | was constructed | was located |
| was controlled | was maintained | was converted | was maximised |
| was created | was measured | was designed | was minimised |
| was derived | was modified | was discarded | was normalised |
| was distributed | was obtained | was operated | was sampled |
| was optimised | was scored | was plotted | was selected |
| was positioned | was separated | was prepared | was simulated |
| was quantified | was stabilised | was recorded | was substituted |
| was regulated | was tracked | was removed | was transferred |
| was repeated | was treated | was restricted | was varied |
| was retained | was utilised | | |

To validate the results from the intrinsic model, samples were collected from all groups.

The method of false nearest neighbours was selected in order to determine the particle dimension.

For the sake of simplicity, only a single value was analysed.

By partitioning the array, all the multipaths could be identified.

Zinc oxide was drawn into the laminate with the intention of enhancing delaminations and cracks.

The advantage of using three-dimensional analysis was that the out-of plane stress field could be obtained.

Because FITC was used for both probes, enumeration was carried out using two different slides.

The LVDTs were unrestrained, so allowing the sample to move freely.

The cylinder was constructed from steel, which avoided problems of water absorption.

最后作者还可以说明这种实验方法可能遇到的问题。任何实验都不是完美的，如果你的实验方法本身具有明显的缺陷，即使这种缺陷不影响实验结果，但如果你不提及，会给读者留下不在意这些内容的坏印象。以下列举一些这部分常用的词汇和示例。

did not align precisely　limited by　only approximate　inevitably
it is recognised that　acceptable　less than ideal　fairly well
not perfect　necessarily　not identical　impractical
slightly problematic　as far as possible　rather time-consuming　(it was) hard to
minor deficit　(it was) difficult to　slightly disappointing　unavoidable
negligible　impossible　unimportant　not possible
immaterial　quite good　a preliminary attempt　reasonably robust
not significant　however　talk about a solution　nevertheless
future work should...　currently in progress　future work will...

Inevitably, considerable computation was involved.

Only a brief observation was feasible, however, given the number in the sample.

Although centrifugation could not remove all the excess solid drug, the amount remaining was negligible.

Solutions using($q=1$) differed slightly from the analytical solutions.

Continuing research will examine a string of dc-dc converters to determine if the predicted efficiencies can be achieved in practice.

While the anode layer was slightly thicker than 13μm, this was a minor deficit.

1.2.5.4 结果与讨论 (Results and Discussion)

不同的期刊对此小节有不同的称呼。不同期刊中该节常与下一节结合组成不同的结构。写作时注意符合期刊的要求。常见的结构见表 1-1。

**表 1-1 常见的结构**

| 类型 1 | 类型 2 | 类型 3 | 类型 4 |
|---|---|---|---|
| Results | Results & Discussion | Results | Results |
| Discussion | — | Discussion & Conclusion | Discussion |
| Conclusion | Conclusion | — | — |

结果与讨论中主要包含实验发现的现象或规律以及针对实验结果的讨论。在实验结果中，可以使用图片、表格或公式等描述实验结果。图表能够清晰简洁地表述大量的信息。列出图表后，还应该有相应的文字说明。文字用于阐述自己的实验现象的理解，说明实验

的创新点或将实验结果与引言中的实验目标结合起来。许多初学者常犯的错误是在该部分中仅仅陈列了实验结果，没有进行解释和进一步的讨论。此外，在讨论时应该进行一定程度的扩展，而不是仅仅局限于实验结果。与引言不同，结果与讨论（Results and Discussion）没有过于严谨的结构。以类型2为例，可以拆解为四个部分。首先重述实验的目的或对实验结果进行总概述。以下列举一些这部分常用的词汇和示例。

as discussed previously
as mentioned earlier/before
as outlined in the introduction
as reported
in order to..., we examined...
it is important to reiterate that...
it is known from the literature that...
it was predicted that...
our aim/purpose/intention was to...
since/because..., we investigated...
the aforementioned theory/aim/prediction etc.
to investigate..., we needed to...
we reasoned/predicted that...
generally speaking
in general
in most/all cases
in the main
in this section, we compare/evaluate/present...
it is apparent that in all/most/the majority of cases
it is evident from the results that...
on the whole
the overall response was...
the results are divided into two parts as follows:
using the method described above, we obtained...

Since the role of copper is critical, the effect of its function during oxidative leaching of refractory gold ore was investigated experimentally.

We reasoned that an interaction between copper and thiosulfate may be essential.

In earlier studies attempts were made to establish degradation rate constants by undertaking ozonation experiments.

The main purpose of this work was to test the smelt performance.

As mentioned previously, the aim of the tests was to construct a continuous separation technology.

In this work, we sought to establish a methodology for the synthesis of a benzoxazine skeleton.

It was suggested in the introduction that the effective stress paths may be used to define local bounding surfaces.

It is apparent that both films exhibit typical mesoporous structures.

It is evident that these results are in good agreement with those reported in references.

In general, coefficients for months close to the mean flowering data were negative.

Our confidence scores have an overall strong concordance with previous predictions.

On the whole, the strains and deflections recorded from the FE model.

Follow similar patterns to those recorded from the vacuum rig tests.

Levels of weight loss were similar in all cases.

然后再描述实验的结果，其中实验的关键结果需要详细的描述，并且附上相应的解释。对于不太重要的结果，可以提也可以不提，提了也不用进行详细的描述。以下列举一些这部分常用的词汇和示例。

(data not shown)

(Fig. 1)

(see also Fig. 1)

(see Fig. 1)

(see Figs. 1-3)

as evident from/in the figure

as illustrated by Fig. 1

as indicated in. Fig. 1

can be seen from/in Figure 1

results are given in Fig. 1

we observe from Fig. 1 that...

accelerate(d) is/are/was/were constant match(ed)

all is/are/was/were different none

change(d) is/are/was/were equal occur(red)

exist(ed) is/are/was/were lower remain(ed) constant

expand(ed) is/are/was/were present remained the same

in particular

in principle

inadequate

strong(ly)

substantial(ly)

sudden(ly)

as anticipated

as expected

compare well with

concur

is/are similar(to)

is/are unlike

The leaching data in Fig. 18 indicate a more reasonable relationship.

The overall volume changes are reported in Fig. 6(d).

Similar results were found after loading additive into the leaching system(data not shown).

Typical cyclic voltammograms for the pyrite can be seen in Fig. 1.

Comparing Figs. 1 and 4 shows that volumetric strains developed after pore pressure had dissipated.

The rate constants shown in Table 1 demonstrate that the reactivity is much greater at neutral pH.

The results are summarized in Table 4.

This kind of delamination did not occur anywhere else.

It eventually leveled off at a terminal velocity of 300m/s.

In the majority of cases, SEM analysis revealed a considerably higher percentage of fine material.

Distributions are almost identical in both cases.

The results are qualitatively similar to those of earlier simulation studies.

These trends are in line with the previously discussed structure of the of the ferrihydrite aggregates.

偏向客观描述的方式有以下词汇和句型。

| | | |
|---|---|---|
| accelerate(d) | match(ed) | is/are/was/were constant |
| all | none | is/are/was/were different |
| change(d) | occur(red) | is/are/was/were equal |
| decline(d) | peak(ed) | is/are/was/were found |
| decrease(d) | precede(d) | is/are/was/were higher |
| delay(ed) | produce(d) | is/are/was/were highest |
| drop(ped) | reduce(d) | is/are/was/were identical |
| exist(ed) | remain(ed) constant | is/are/was/were lower |
| expand(ed) | remained the same | is/are/was/were present |
| fall/fell | rise/rose | is/are/was/were seen |
| find/found | sole(ly) | is/are/was/were unaffected |
| increase(d) | vary/varied | is/are/was/were unchanged |

偏向主观描述的方式有以下词汇和句型。

| | | | |
|---|---|---|---|
| abundant(ly) | imperceptible(ibly) | remarkable(ably) | acceptable(ably) |
| important(ly) | resembling | adequate(ly) | in particular |
| satisfactory | almost | in principle | scarce(ly) |
| appreciable(ably) | inadequate | serious(ly) | appropriate(ly) |
| interesting(ly) | severe(ly) | brief(ly) | it appears that |
| sharp(ly) | clear(ly) | large(ly) | significant(ly) |
| comparable(ably) | likelihood | similar | considerable(ably) |
| low | simple(ply) | consistent(ly) | main(ly) |
| smooth(ly) | distinct(ly) | marked(ly) | somewhat |

dominant(ly)
mild(ly)
strong(ly)
essential(ly)
negligible( ibly)
suitable( ably)
extensive(ly)
overwhelming(ly)
too+adjective
general(ly)
radical(ly)
very

measurable( ably)
striking(ly)
equivalent
most(ly)
sufficient(ly)
exceptional(ly)
only
the majority of
few
quick(ly)
valuable
immense(ly)

steep(ly)
drastic( ally)
more or less
sudden(ly)
excessive(ly)
obvious(ly)
tendency
fair(ly)
powerful(ly)
unusual(ly)
high(ly)
virtual(ly)

dramatic( ally)
minimal(ly)
substantial(ly)
excellent
noticeable( ably)
surprising(ly)
extreme(ly)
poor(ly)
unexpected(ly)
good
rapid(ly)

此外，还可以与其他学者的实验结果进行对比，以此说明自己实验结果的优点或不足。以下列举一些这部分常用的词汇和示例。

as anticipated
as predicted by…
compare well with
confirm
contrary to
correlate
inconsistent with

is/are better than
is/are identical(to)
is/are parallel(to)
is/are unlike
prove
reinforce
validate

as expected
as reported by…
concur
consistent with
corroborate
disprove
in line with

is/are in good agreement
is/are not dissimilar(to)
is/are similar(to)
match
refute
support
verify

It is evident that the extraction results obtained here are in exceptionally good agreement with existing FE results.

Distributions are almost identical in both cases.

Our concordance scores strongly confirm previous predictions.

We see that the numerical model tends to give predictions that are parallel to the experimental data from corresponding tests.

These results demonstrate that improved correlation with the experimental results was achieved using the new mesh.

The results are qualitatively similar to those of earlier simulation studies.

These trends are in line with the previously discussed structure of the of the ferrihydrite aggregates.

其次，如果实验结果有缺陷或问题，也需要提出来进行讨论或解释。以下列举一些这部分常用的词汇和示例。

(a) preliminary attempt
despite this
however
immaterial

may/could/might have been
or
was/were
beyond the scope of this study

incomplete
infinitesimal
insignificant
less than ideal
less than perfect
(a) minor deficit/limitation
negligible
nevertheless
not always reliable
not always accurate
not ideal
not identical
not completely clear
not perfect
not precise
not significant
of no consequence
of no/little significance
only
reasonable results were obtained
room for improvement
slightly( disappointing)
(a) slight mismatch/limitation
somewhat( problematic)
(a) technicality
caused by
difficult to( simulate)
due to
hard to( control)
inevitable
it should be noted that...
not attempted
not examined
not explored in this study
not investigated
not the focus of this paper
not within the scope of this study
possible source(s) of error
unavoidable
unexpected
unfortunately
unpredictable
unworkable
unavailable
further work is planned
future work should...
future work will...
in future, care should be taken
in future, it is advised that...
unimportant

The correlation between the two methods was somewhat less in the case of a central concentrated point load.

It should, however, be noted that in the ANSY methods, the degree of mesh refinement may affect the results.

Nevertheless, this effect is only local.

Full experimental data was only obtained for one local ore sample.

Reasonable results were obtained in the first case, and good results in the second.

It is difficult to simulate the behavior of nickel migration during leaching.

Although this was not obtained experimentally, it can be assumed to exist.

Future work should therefore include numerical diffusion effects in the calculation of permeability.

This type of control saturation is fairly common and therefore of no significance.

最后，如果实验结果有其他可能的解释，也可以在最后说明。以下列举一些这部分常用的词汇和示例。

apparently
it is logical that

could* be due to
could* be explained by
could* account for
could* be attributed to
could* be interpreted as
could* be seen as
evidently
imply/implies that
indicate/indicating that
in some circumstances
is owing to
is/are associated with
is/are likely
is/are linked to
is/are related to
it appears that
it could* be concluded that…
it could* be inferred that
it could* be speculated that
it could* be assumed that
it is conceivable that
it is evident that
it is thought/believed that
it seems that
it seems plausible(etc.) that
likely
may/might
means that
perhaps
possibly/possibility
potentially
presumably
probably
provide compelling evidence
seem to
suggest(ing) that
support the idea that
tend to
tendency
unlikely
there is evidence for
we could* infer that
we have confidence that
would seem to suggest/indicate

注：could* 表示该词可以被 may, might, can 替换。

This suggests that silicon is intrinsically involved in the precipitation mechanism.

These curves indicate that the effective breadth is a minimum at the point of application of the load.

Empirically, it seems that alignment is most sensitive to rotation in depth.

Only the addition of choride produced a positive response, suggesting that other species in the system exhibit little effect.

It is therefore speculated that at pH 7.5 a major part of the reaction was via hydroxyl radical attack.

It is apparent that this type of controller may be more sensitive to plant/model mismatch than was assumed in simulation studies.

The results seem to indicate that this causes the behavior to become extremely volatile.

#### 1.2.5.5 结论（Conclusion）

结论部分不是必须要有的。以期刊“Metals”为例，对结论（Conclusion）部分的要求如下：

This section is not mandatory but can be added to the manuscript if the discussion is unusually long or complex.

通常而言，当结果与讨论部分如果比较长且复杂时，可以在结论中进行概括性的总结，该部分没有固定的结构，作者根据文章内容写作即可。下列举一些这部分常用的词汇和示例。

accurate
advantage
appropriate
assist
attractive
beneficial
better
clear
compare well with
compelling
comprehensive
confirm
convenient
convincing
correct
cost-effective
could lead to
crucial
dramatic
easy
efficient
effective
enable
encouraging
enhance
ensure
evident
exact
excellent
exceptional
exciting
extraordinary
facilitate
feasible
flexible
help to
ideal
important
improve
invaluable
is able to
low-cost
novel
offer an understanding of
outperform
outstanding
overwhelming
perfect
powerful
productive
prove
provide a framework
provide insight into
provide the first evidence
realistic
relevant
remarkable
remove the need for
represent a new approach to
reveal
robust
rule out
simple
solve
stable
straightforward
striking
strong
succeed in
successful
superior
support
surprising
undeniable
undeniable
unique
unprecedented
unusual
useful
valid
valuable
vital
yield

The presence of such high levels of copper is a novel finding.

We identify dramatically different profiles in catalytic effect of nickel.

Our results provide compelling evidence that this facilitated gold extraction.

These preliminary results demonstrate the feasibility of using hydroperoxide as the reductive reagent.

Our data rule out the possibility that this behaviour was a result of metal precipitation.

The system presented here is a cost-effective detection protocol.

A straightforward analysis procedure was presented which enables the accurate prediction of metal leaching behavior.

Our study provides the framework for future studies to assess the performance characteristics.

We have made the surprising observation that the extraction of cobalt is also pH dependent.

We have derived exact analytic expressions for the percolation threshold.

Our results provide a clear distinction between the functions of the pathway metal ions.

### 1.2.6 致谢部分

致谢（Acknowledgement）不是所有的期刊都有要求，但如果论文中的内容涉及别人的帮助、资助等，则应该提出致谢。因此这一部分是表示作者对某人或某单位的感谢。表示感谢的词汇有以下几类：

| | | | |
|---|---|---|---|
| acknowledge | grateful | recognition | acknowledgment |
| gratitude | thank | appreciate | indebted |
| thankful | appreciation | indebtedness | thanks |

此外，还有以下句型可以使用：

上述各项中的名词+go to/be due to+人/be owed to

还有一些表示感谢的句型可直接使用：

I'd like to thank

The author wishes to express his sincere thanks to...

I'd like to acknowledge

The author wishes to make the acknowledgement of...

Grateful acknowledgement is made to...

The author is in acknowledgement of...

I acknowledge...

I would like to express my gratitude to...

Special gratitude is owed to...

A special gratitude is expressed to...

It is a pleasure to record my gratitude to...

I'm grateful to...

I greatly appreciate...

I would like to express my appreciation of...

I'm greatly indebted to...

I'm in indebtedness to...

We express our indebtedness to...

I wish to thank...

Thanks are owed to...

Thanks go to...

Thanks must be given to...

Thanks must be extended to...

Special thanks go to...

I offer my thanks to...

A word of thanks is given to...

Particular recognition is due to...

We wish to give special thanks to...

His sincere thanks are offered to...

We are thankful to...

We wish to record our appreciation to...

With a great sense of debt, we thank...

作者为了避免语言上的死板、乏味，在同一篇论文的致谢项目中若有几个表示感谢的句子的话，则一般使用上面所提到的不同句型。

### 1.2.7 参考文献

大多数国外期刊对参考文献（References）的要求都是在集中在最后的“Reference”章节里，并按序号排列，也有部分期刊要求把参考文献分别列在每一页的脚注中。

1.2.7.1 参考书的一般书写顺序

各种期刊论文中参考书的书写顺序大致是相同的，仅某些项目前后顺序略有不同，下面介绍两种写法。

第一种写法如下：

作者姓名．书名（通常用斜体表示，也可用引号标出）．出版社地点：出版社名称，出版年份，页码（也可不标出此项）.

例如：

[1] R. W. Brockett. *Finite Dimensional Linear Systems*. New York: Wiley, 1970.

[2] V. Belevithch. *Classical Network Theory*. San Francisco, CA: Holden-Day, 1968.

[3] D. E. Dudgeon, R. M. Mersereau. *Multidimensional Digital Signal Processing*. Englewood Cliffs, NJ: Prentice-Hall, 1984.

第二种写法如下：

作者姓名，书名，出版社名称，出版社地点，第几章或第几节和页码，出版年份.

[1] J. R. Pierce, “Theory and Design of Electron Beams”, D. Van Nostrand Co., Inc., New York, N. Y., 2nd ed., ch. 6, pp. 72-91; 1954.

[2] V. K. Zworykin, et al., “Electron Optics and the Electron Microscope”, John Wiley & Sons, Inc., New York, N. Y., sec. 10. 5, p. 355; 1945.

1.2.7.2 参考期刊文章的一般书写顺序

作者姓名，文章标题（通常用引号标出），杂志名称（一般用斜体）.，卷号和页码，刊登的月份和年份.

需要注意的是，如果参考的是即将要发表的文章，刊登日期可以写为 to be published（待发表）字样；也有写上 in press（正在排印）或 preprint（予印本：指因某种需要而自行付印的单行本）的字样。例如：

[1] H. J. Orchard, “Inductorless filters”, *Electron Lett.*, Vol. 2, pp. 224-225, June 1966.

[2] A, L. Schawlow, “Infra-red and optical masers”, *Solid state J.*, Vol. 2, pp. 21-27;

June 1961.

[3] A. L. Schawlow, "Finite structure and properties of chromium fluorescence in aluminum and magnesium oxide", *Advances in Quantum Electronics*, to be published.

### 1.2.8 作者简介

1.2.8.1 作者简介内容

一般期刊不需要作者简介这一部分。但有些著名期刊在论文结束后还要有作者简介，有的还需要作者的照片（如IEEE），所以作者本人应把这一简介的英文文本随同论文稿件一并寄上。作者简介一般需要提到以下几个方面的内容：

（1）自大学起获得的学位名称、时间、专业和学校及其地点。

（2）工作经历（包括时间、职位、工作单位及地点）。

（3）目前工作的侧重点或科研的兴趣及方向。

（4）突出的论文和主题、科研成果、重点工程及获奖情况等。如果是著名学会的成员，则也要写上。

1.2.8.2 常用句型

（1）用于学历方面的句型：

He attended… University from 19×× to 20××, majoring in…

从19××年至20××年在××大学学习××专业。

（2）用于获得学位的句型：

In 20××, he/she received/obtained/earned/was awarded a… degree in… from… Department at… University, … city/state/province/country

在20××年在××地方的××大学××系获得了××学位。

（3）主要学位的名称

在国际期刊上常见的学位名称有：

| | | | |
|---|---|---|---|
| B. S. /B. Sc. | 理学士 | B. A. | 文学士 |
| B. Eng. | 工学士 | M. S. | 理硕士 |
| M. A. | 文硕士 | M. Eng. | 工硕士 |
| Ph. D. | （哲学）博士 | Doctor of Engineering | 工程（学）博士 |

1.2.8.3 用于工作经历的句型

（1）在……（机构）担任……（职位）。

He works/acts/serves as… in/at/for…

He is/was/has been with/at… as…

He is employed in the position of…for/at…

He holds the position of…for/at…

（2）是××大学××系××专业的副教授。

He is(an) associate/adjunct professor of…(speciality) in… Department at… University, …(place).

(3) 是××大学的教师。

He is/has been a faculty member of... University.

(4) 成为××大学的教师。

He joined the faculty of... University.

(5) 在××大学攻读××专业博士学位。

He has been working towards/on the Ph. D. Degree in...(speciality) at... University.

He is currently/presently working towards/on the Ph. D. Degree in...(speciality) at... University.

#### 1.2.8.4 用于科研方面的句型

His research interests/work focus focus(es)/concentrate(s) on the area(s)/field(s) of...

His research interests/work include(s)/concern(s)/are/is in the area(s)/field(s) of...

He is actively engaged in(research in the area(s) of) ...

He has been engaged in(research in the area(s) of) ...

He is active in(the area(s) of) ...

#### 1.2.8.5 用于科研成果及获奖方面的句型

He invented...

He has made contributions specifically in...

He received the medal/award of...

He was awarded...

He is the recipient of...award.

He has published... books and... papers.

He is(the) author/coauthor of the book...

A 2019 paper was the winner of... award.

A 2019 paper won... award.

# 科技英语写作词汇与使用

## 2.1 词汇使用原则

在撰写有色冶金英语论文时，有时词汇的选择能够决定了这篇论文阅读的难易程度。如果能够选择合适的词汇，将有助于学术的交流。写作的目的是交流信息，通常我们可以通过选择合适的词汇达到这一目的。本节将介绍一些有色冶金科技论文写作词汇选择的基本原则。

### 2.1.1 词汇的词源

学术写作与日常英语具有较大差别，在使用时需要注意区分。学术写作词汇的词源常为希腊语和拉丁语，主要原因是拉丁语是欧洲文艺复兴时期使用的学术语言。当进行冶金科技英语写作时，用于描述现象和概念的词汇的词源常为希腊语和拉丁语。写作时应注意区分日常英语词汇和学术写作时使用的词汇。以下列举一些初学者常常使用不恰当的科技英语写作词汇。

| 日常英语 | 学术写作 | 日常英语 | 学术写作 |
|---|---|---|---|
| worry | concern | story | account |
| get rid of | eradicate | a lot of | considerable |
| not enough | insufficient | trouble | difficulty |
| big | significant | way | method |
| bring together | synthesize | thing | object |
| about | approximately | ask | inquire |
| begin | commence | buy | purchase |
| carry | bear | change | transform |
| cheap | inexpensive | end | conclude |
| finish | complete | get | obtain |
| give | accord | have | possess |
| method | technique | obtainable | available |
| avoid | circumvent | quick | rapid |
| say | remark | similar | identical |
| touch | contact | try | endeavor |
| use | employ 或 utilize | put | place |
| fire | flame | happy | excited |
| careful | cautious | care | caution |
| heart | center | enough | sufficient |
| by which | whereby | in which | wherein |
| in the end | eventually | | |

### 2.1.2 避免使用拖沓、重复的表述

一些短语完全可以用一个词语代替，进而使得文章简洁明了。

| | | | |
|---|---|---|---|
| the reason is that/because | because | it was evident that | apparently |

一些短语动词往往由单独动词替代，表现出科技英语行文要求精炼，表达上力求简洁。

| 短语动词 | 单独的动词 | 短语动词 | 单独的动词 |
|---|---|---|---|
| turn on | complete | take in | absorb |
| push in | insert | push down | depress |
| put up | erect | put together | aggregate |
| put out | extinguish | put in | add |
| wear away | erode | take away | remove |
| send for | summon | use up | consume, exhaust |
| carry out | perform | come across | encounter |
| make up | invent | put up with | tolerate |
| go with | accompany | fill up | occupy |
| bring out | introduce | break up | rupture |
| look at | examine | find out | discover |
| drive forward | propel | keep up | maintain |
| turn upside down | invert | set fire to | ignite |
| make… weak | weaken | take… into pieces | dismantle |
| turn… into liquid | liquefy | be made up of | be composed of |

### 2.1.3 避免使用主观性太强的抽象词汇

当描述一些实验现象或实验结果时，使用主观性太强的抽象词汇会导致论文阅读起来失去客观性，写作时应尽量避免。

| | | | |
|---|---|---|---|
| awesome | brilliant | extraordinary | excellent |

### 2.1.4 同义词的使用

当某一词汇需要重复使用时，可以使用同义词替换，使文章看起来丰富不单调。需要引用数据时，除了 show，present 等外，还可以使用以下词汇表述。

| | | | |
|---|---|---|---|
| illustrate | summarize | demonstrate | contain |
| list | report | reveal | display |
| describe | suggest | | |

引用他人的研究结论/观点：

state claim argue maintain
suggest assert conclude describe
examine presume

## 2.2 动　　词

与中文不同，英语中的动词具有表达时态、被动和虚拟语气等功能。关于这些语法内容，市场上的语法书有更清晰的表述，本书在此不再赘述。这里将会讨论一些略微复杂的语法规则，这些规则在科技英语写作中需要注意。

### 2.2.1 主谓一致

一个句子的谓语需要和主语的单复数形式一致。这条规则很容易理解，但是在实际写作时却非常容易出错。其原因在于写作时对于主语单复数的判断不一定准确，其次，有时句子中哪个是主语也不容易判断。一般遇到不容易判断主语的句子时，可以删去句子中的定语、状语和补语，只留下主语、谓语和宾语，再进行判断就容易得多了。

Results of the high-temperature experiment are described.

如例句中，删除修饰部分后，句子的核心内容就是

Results are described.

很容易判断句子的主语是 results 而不是 experiment，因此谓语应该用 are 而不是 is。

### 2.2.2 动词语态

语态可分为主动语态和被动语态两类。在科技论文写作中，常使用被动语态表现论文内容的客观性。例如，当研究完某个课题后，一般不会说，“I investigated ×××” 而会使用被动语态来表述这个行为：

××× was investigated.

通过这样的表述，消除了表达中的“我”，从而使语句的表述更加客观。

一些科技英语写作中常用动词的不规则变化需要了解，本节在此总结了一些供大家参考，见表 2-1。

**表 2-1　常用动词的不规则变化**

| 原式 | 过去式 | 过去分词 | 原式 | 过去式 | 过去分词 |
|---|---|---|---|---|---|
| set | set | set | understand | understood | understood |
| cut | cut | cut | find | found | found |
| cost | cost | cost | become | became | become |
| hit | hit | hit | begin | began | begun |
| let | let | let | break | broke | broken |
| shut | shut | shut | choose | chose | chosen |
| say | said | said | rise | rose | risen |

续表 2-1

| 原式 | 过去式 | 过去分词 | 原式 | 过去式 | 过去分词 |
|---|---|---|---|---|---|
| get | got | got | write | wrote | written |
| sit | sat | sat | take | took | taken |
| spend | spent | spent | know | knew | known |
| think | thought | thought | blow | blew | blown |
| make | made | made | do | did | done |
| bring | brought | brought | fall | fell | fallen |
| smell | smelt | smelt | give | gave | given |
| lead | led | led | show | showed | shown |

### 2.2.3　特殊连系动词的使用

特殊连系动词是指由少数实义动词变成的连系动词，如 get, turn, go, stay, look 等。

This figure looks puzzling.

这图看起来令人费解。

Its result proves correct.

其结果证明是正确的。

In this case, the output of anodic current stays stable.

在这种情况下，阳极输出电流保持稳定。

## 2.3　动词的非谓语形式

### 2.3.1　动词不定式

在冶金科技英语写作中应该注意以下几点：

(1) 动词不定式作主语时且使用形式主语“it”时最常用的几个句型：

1) ……能够……

It is possible to(do…).

It is now possible to convert the equation into…

现在能够把这个方程转化为……

It seems possible to solve this equation.

似乎能够解这个方程。

2) ……必须……

It is necessary to(do…).

It is necessary(for us) to have a good command of Kirchhoff's laws.

我们必须掌握好基尔霍夫定律。

It will not be necessary to calculate the extent of the leaner function for leaching nickel and cobalt.

不必计算镍钴浸出的线性关系。

3）……需要（花费）……

It takes/requires... to( do...).

It takes about eight minutes for the lixvant to react with gold.

浸出剂和金反应需要大约 8 分钟。

It takes about 20 gram of copper to precipitate 30 gram of copper sulfide.

产出 30 克铜硫化物沉淀，需要大约 20 克铜。

We can find how many calories it will require to melt the ice.

我们能求出，要融化这些冰需要多少卡的能量。

4）……想要……

It is desired to( do...).

If it is desired to increase the heat flow, an increase in the hS product is required.

若想要增加热流，就需要增大 hS 积。

（2）作定语的几个句型：

1）"介词+which+动词不定式"的句型（这时被修饰的名词为普通名词，且该名词不定式之间并不存在主谓关系或动宾关系）。

Each element have different atomic mass with which to indicate its property.

每种元素的不同的元素质量决定了它的性质。

Aluminum can be taken as the light metal which to include.

铝可作为轻金属中的一种。

We shall use such a field on which to base our discussion of metal precipitation.

我们将使用这种场作为讨论金属沉淀的基础。

2）the ability（tendency，capacity，capability，desire，failure 等）+of A to do B 的句型。

Energy is defined as the ability of a body to do work.

能量被定义为物体做功的能力。

The desire of non-ferrous metal to obtain better elastic has been the goal in some fields.

使得有色金属获得更好的弹性成为某些领域的目标。

The greater the tendency of an object to resist a change of velocity, the greater its inertia.

物体阻止速度变化的趋势越大，其惯性就越大。

The capacity of air to absorb water vapor increases as its temperature rises.

空气吸收水蒸气的能力随温度的上升而提高。

The deviations from the expected periodicity in Mendeev's list were due to failure of contemporary chemistry to have discovered some of the elements existing in nature.

那时之所以与门捷列夫周期表中所预料的周期性有出入，是由于当时化学界未能发现存在于自然界的某些元素之故。

（3）其他几个常用句型。

1）主语+系动词+形容词+不定式主动形式：

This problem is very difficult to solve.

这道题很难解。

This case is particularly simple to analyze.

这种情况分析起来特别简单。

The test of leaching and recovery of metal is easy to operate.

金属浸出和回收的实验很容易操作。

The methods for these tests are essentially as easy to operate as precious ones.

这些实验使用的方法操作起来基本上与前述的一样容易。

2）主语+及物动词+it+形容词+不定式短语：

We find it very easy to solve this problem.

我们发现解这道题是很容易的。

We consider it necessary to reduce pollution in metallurgical engineering.

我们认为必须减少冶金工程中的污染。

The invention of Hall-Héroult process made it possible to find out a cheaper way to acquire aluminum.

霍尔-埃鲁铝电解法的发明使得人们获得了更便宜的制造铝的方法。

3）主语+及物动词+名词+形容词+不定式主动形式：

We find the theory of copper leaching kinetics in choride solution very difficult to explain.

我们发现铜在氯化物溶液中的浸出动力学是很难解释的。

We find the leaching rate of gold difficult to measure.

我们发现金的浸出速率难以测量。

The author wishes to thank the publishers for their assistance throughout and for being so easy to work with.

作者要感谢出版社在本书整个出版过程中所给予的帮助以及极为融洽的合作。

4）名词性不定式一般应该使用主动形式：

We have now to consider what way of metal metallurgy to take.

我们现在必须考虑采取哪一种冶金工艺。

We have to find how many sodium hydroxide to add so as for crocoite to decompose.

我们得知道要加入多少氢氧化钠才能使赤铅矿分解。

（4）不定式标志“to”可省略的场合。当句子的主语部分含有实意动词“do”的任何形式（无论是谓语还是非谓语形式）时，作句子表语的动词不定式的标志“to”可以省略。如：

What a generator dose is convert mechanical energy into electrical energy.

发动机的用途是把机械能转换为电能。

The best way to do it is remove the smelt slag.

对它最好的处理办法是，把熔融渣移走。

### 2.3.2 分词

#### 2.3.2.1 分词作定语的情况

分词作定语一般遵循“单分在前，分短在后”的原则，即单个分词作定语时一般处于被修饰词的前面，而分词短语作定语时一定要放在被修饰词之后。写作的重点放在分词短语作后置定语的情况。

A charge body can attract the diffusion of copper ions.

带电体能吸引铜离子的扩散。

These moving electrons facilitate the migration of nickel ions.

这些运动的电子促进了镍离子的迁移。

The gauss is a very commonly used unit for the measurement of flux density.

高斯是测量磁通量密度的一个很常用的单位。

Students using this book have completed a course in Non-ferrous Metallurgy.

使用本书的学生，已经学完了有色金属冶金学课程。

In most cases, the greater the temperature, the greater leaching rate obtained in non-ferrous metallurgy.

多数情况下，温度越高，在有色金属冶金工艺中的浸出率也越高。

The free electrons in a conductor carrying charge also play an important role in the electro-deposition process.

在导体中携带电荷的自由电子在电沉积过程中起了重要的作用。

A cathode usually is a device consisting of inert metal such as platinum or silver in electrolytic process.

在电解过程中，阴极常为铂、银等惰性金属组成的电极。

The greater the temperature, the greater the leaching rate.

外力越大，加速度就越大。

Hydrogen is the lightest element known.

氢是人们所知的最轻的元素。

The assumptions made will affect the type of series obtained.

所作的假设会影响获得的级数的类型。

Finally, some technical problems remaining are outlined.

最后，概述了有待于解决的一些技术问题。

2.3.2.2 分词作状语的情况

写作时应侧重于句首（主要表示时间、条件、原因及对主语的附加说明）和句尾（主要表示伴随状况、结果及方式等）的情况，这是典型的书面语言形式，使句子结构紧凑、简洁，可许多读者不会使用这一结构。

Being a good conductor of electricity, copper is widely used in electrical engineering.

由于铜是良导体，因此被广泛地应用在电气工程中。

Flowing through a circuit, the current will lose part of its energy.

当电流流过电路时，要损耗掉一部分能量。

Given current and resistance, we can find out voltage.

若已知电流和电阻，我们就能求出电压来。

Lacking knowledge of just what these radiations were, the experimenters named them simply alpha, beta, and gamma radiation, from the first three letters of the Greek alphabet.

因为这些实验者当时并不知道这些射线到底是什么东西，所以他们就按照希腊字母表的头三个字母把它们分别命名为阿尔法射线、贝塔射线和伽马射线。

Having obtained the initial conditions, we go on to do pre-experimental.

在获得了初始条件后，我们接下去就要做预实验了。

#### 2.3.2.3 分词独立结构

分词独立结构同样是典型的书面语言形式，写作重点放在句尾时作附加说明的情况。

This current changing, the rate of metal precipitation will change as well.

该电流变化时，金属沉淀速率也将发生变化。

A power reactor having no need of air, we can build it underground.

由于电力反应堆不需要空气，因此我们可以把它建在地下。

This done, the function for copper leaching becomes much simpler.

这样处理之后，铜浸出公式变得简单多了。

Almost all metals are good conductors, sliver being the best.

几乎所有的金属都是良导体，而银为最好。

The sign of the integral depends on the direction of the path taken, a counter-clockwise direction being taken as positive.

积分的符号取决于所取路径的方向，我们把逆时针方向取为正。

There are several basic laws governing these interactions, all of them discovered early in the nineteenth century.

支配这些相互作用的基本定律有几个，它们都是在19世纪初发现的。

The tube floats vertically in the leaching vessel, heavy end down.

该管子垂直地浮在浸出容器中，其重的一头朝下。

The force can be resolved into two components, one of them vertical and the other horizontal.

这个力可以被分解成两个分量，一个是垂直的，而另一个是水平的。

#### 2.3.2.4 with 和 without 结构

所谓“with结构”指的是以下这一形式：with+名词/代词+分词（短语）/介词短语/形容词（短语）/副词/不定式短语/名词（短语）。

在科技写作中已广泛使用这种结构（“without”可用来引出这一结构的否定形式；也可用“with+no/neither/none…”来表示否定含义），读者应熟练掌握以下两点：

（1）作状语：写作重点放在处于句首的表示条件、时间和原因；处于句尾表示附加说明、方式、条件等。

1）在句首的情况：

With the development of its base function, the catalytic effect of calcium is significant.

在其基本功能的作用下，钙的催化效果显著。

With the circuit open, the voltage between anode and cathode equal to the balance voltage.

当电路打开时，阴阳极间电位等于平衡电压。

With its stirring speed 0, stir will be cut off.

若搅拌速度为零，搅拌就停止了。

With this in view, we have attempted to write this book.

由于考虑到了这一点，我们才力图编写本书的。

Without the temperature or pH changed, ore can never be leached.

如果温度或 pH 值不发生变化，矿石是永远不能被浸出的。

Without the air to stop some of the sun's heat, every part of the earth would be burning hot when the sun's rays strike it.

要不是空气挡住了一部分太阳的热量，在阳光照射时，地球的任何一部分都会灼热不堪。

With the leaching systems becoming more complex, the achievement of reliability has become an increasingly difficult problem.

由于浸出系统越来越复杂，实现可靠性已成为一个越来越困难的问题。

With these definitions in hand, Eq. (6) can be transformed term by term.

有了这些定义后，就可对式（6）一项一项地进行变换了。

2）在句尾的情况：

Both practical design techniques and theoretical problems are covered with emphasis on general concepts for solvent extration.

本书既讲了实际的设计方法，同时也讲述了理论问题，而侧重点则放在溶剂萃取的一般概念上。

A positively charged ring of radius $R$ is placed with its plane perpendicular to the $x$-axis and with its center at the origin.

半径为 $R$ 的带正电的环境被放置成其平面垂直于 $x$ 轴，而其圆心在原点。

It is impossible for heat to be converted into a certain energy without something lost.

使热转变成某种能量而不损失什么东西是不可能的。

Some pyrometallurgy was not designed with environmental protection in mind.

一些火法冶金的流程设计时没有考虑环保问题。

（2）作定语的情况：

This is a triangle with the apex up.

这是一个顶点朝上的三角形。

The earth is perhaps the only planet with life on it.

地球也许是在上面有生命存在的唯一行星。

This is the most effective leaching way with low power consumption.

这是一个效率最高的浸出方法并且能耗很低。

### 2.3.3　动名词

#### 2.3.3.1　动名词作主语

动名词作主语表示一般的情况，不涉及发出动作的具体时间，这是与动词不定式作主语时的主要区别。

Selecting a desired extractant is only one of three important functions performed by the solvent extraction process.

选择所需的萃取剂只是萃取过程的三个重要功能之一。

Proper installing the operating facilitates increases the accuracy of the leaching results.

正确安装操作部件提高浸出实验的准确性。

It is worth noting that a lens has two focal points, one on each side of the lens the distance $f$ from its center.

值得注意的是，一个透镜有两个焦点，它们分别在透镜的两边离其中心 $f$ 这一距离处。

2.3.3.2 动名词作宾语

在某些动词后要表示动作的话，必须使用动名词。在科技写作中常见的这类动词有 avoid, consider, involve, facilitate, require, mean, finish, suggest, practice 等。如：

Let us consider designing a metal separation process.

让我们考虑设计一个金属分离技术。

It is essential to avoid making excessive pollution of the leaching process.

必须避免在浸出过程中产生过多的污染。

Inverse feedback involves taking a small percentage of the output and feeding it back out of phase to a preceding stage.

负反馈就是取出一小部分输出，并把它反馈回前一级。

These problems require finding a most effective way to leach gold.

这里的问题是找出浸金的最有效的方法。

2.3.3.3 动名词的被动形式

In this case, the Na and Cl nuclei cease being shielded by their electrons.

在这种情况下，钠和氯原子核不再由它们的电子所屏蔽。

Ore is capable of being broken down into various components.

矿石能够被分解成各种成分。

Some materials are known as elastic because they return to their original shape after having been bent.

有些材料被称为弹性体，因为它们被弯曲后仍能恢复原状。

2.3.3.4 其他表达法的含义

如“by+动名词”译为“通过……”，它一般可以用“through+名词”来代替；“on/upon+动名词”译为“一……就……”“在…… 之后”“在……时候”；“in+动名词”译为“在……时候/期间”“在……方面”。

By analyzing the performance of the device, one can appreciate it better.

通过分析该设备的性能，我们能更好地了解它。

By finding two points on the straight line, we can draw the line.

通过找到直线上的两点，我们就能把这直线画出来。

On being raised temperature, the leaching rate of iron will be improved.

温度一旦升高，铁的浸出率就会提高。

The energy lost in causing the current to pass through resistance is transformed into heat.

电流通过电阻的过程中所消耗的能量，被转换成了热。

In using this equation, we must pay attention to the sign.

在使用这个式子时，我们必须要注意符号。

The following table may be helpful in converting measurements to different units.

下表在把测量的结果转换成不同单位方面可能是有帮助的。

2.3.3.5 动名词复合结构

通常动名词的复合结构构成如下：物主代词/名词所有格/名词普通格+动名词。其主要功能与普通动名词相同，可以做介词宾语、主语、宾语或表语等，如：

A process circuit can be modified by inserting a leaching circuit in practice.

工艺流程在实践中可通过插入一个浸出流程加以修改。

We know of the earth's acting as big magnet.

我们知道地球的作用就像一块大磁铁。

Automation is not a question of machines replacing man.

自动化并不是机器替代人的问题。

The regular array of atoms in the lattice results in there being certain sets of parallel and equally spaced planes in the crystal.

原子在晶格中的规则排列，促使在晶体中形成了几组相互平行且等间隔的平面。

Like poles repelling each other is a very useful physical phenomenon.

同性磁极相互排斥，是一种很有用的物理现象。

## 2.4 形容词

形容词通常是用于修饰名词的词。在英语的表述中，名词除了用单独的形容词修饰外，还可以用不定式、定语从句或动名词等进行修饰，例如：

| | |
|---|---|
| There is question to be answered. | 不定式 to be answered 修饰名词 question |
| There is question which should be answered. | 定语从句 which should be answered 修饰名词 question |
| The researcher raised a confusing conclusion. | 动名词 confusing 修饰名词 conclusion |

形容词的位置一般位于被修饰名词的附近，其中单独的形容词放置于名词之前，而带有形容词性质的短语（不定式，定语从句或动名词等）一般位于被修饰名词的后面。以下分情况举例说明。

### 2.4.1 不定代词后的形容词

不定代词如 some, every, any, no 与 thing, body, one 的组合。

Now there is nothing mysterious about solvent extraction.

现在萃取技术没什么神秘的了。

This book contains something new.

这本书里有些新内容。

Everything metallurgy will be designed environmentally.

所有的冶金工艺流程设计时都要考虑环保。

### 2.4.2 形容词作后置定语

（1）有一些形容词按语法规则需要后置，如 present（存在的），else（其他的，别的），whatever（任何的）等。

This charge interacts with charges of other species present in the leaching system.

这个电荷与浸出系统存在的其他物质的电荷相互作用。

（2）一些形容词可以放在被修饰的名词后，如 available, obtainable, achievable, necessary 等。

There are the smallest particles obtainable through the leaching processes.

这些是通过浸出过程所能获得的最小颗粒。

This measure is the key for the more effective metallurgy process achievable.

这种测量在达到更高效的冶金工艺中十分重要。

（3）由“and”或“both...and”以及“or”或“either...or”连接的两个形容词作后置定语。

Every particle in the leaching vessel, large or small, possesses gravitation.

浸出容器中每个颗粒，无论大小，都具有重力。

This book is a help to designers of metallurgy process both new and old.

这本书对新老冶金流程设计人员都是有用的。

The power rule can be used for all rational exponents, positive and negative.

这个幂的规则能用于一切有理数，包括正的和负的。

The extration behavior of copper is influenced by either the concentration of chlorie or sulfide.

铜的提取行为受氯化物或者硫化物浓度的影响。

### 2.4.3 形容词短语

（1）作后置定语：

The electrons able to move freely within the ore particles play an extremely important role in the formation of electric current.

在矿石颗粒内能自由运动的电子在形成电流方面起了极为重要的作用。

Infinity is a quantity greater than any number.

无穷大是大于任何数的一个量。

An example of this is motion of chloride ion in sodium choride solution.

这种情况的一个例子是氯化钠溶液中氯离子的运动。

（2）作状语。处于句首时表示主要原因，还可表示条件、让步、对主语的附加说明或对全局的评注性状语等。

Simple in structure and low in price, this device is in great demand.

这种设备由于结构简单、价格低廉，因此需求量很大。

Accurate in operation and high in speed, the automation process can save man a lot of time labor in metallurgy.

由于自动化操作准确且速度快，因此在冶金工业中能节省人类大量的时间和劳力。

## 2.5 名　　词

### 2.5.1 动词的名词化

科技英语写作也会随着时间出现一些风格的变化。目前有一种学术写作的趋势是将描述现象的动词名词化。动词名词化后，句子的主语常变为较长的短语。如：

原：The emission rate of mercury increased sharply.

改：The increasement of emission rate of mercury occurred sharply.

原：Unwanted Roman children were generally abandoned in a public place.

改：The abandonment of unwanted Roman children generally occurred in a public place.

第一句中将动词 increased 名词化为 increasement，第二句中将 abandoned 名词化为 abandonment。这种写作风格需要注意。

### 2.5.2 名词作状语

Small leaching equipment is often lead acid battery powered.

小型浸出设备往往由铅酸电池供电。

The local diffusion is phase stationed to the counter current.

原位扩散相位固定在反向电流。

### 2.5.3 名词短语作同位语

名词短语可作整个句子或句子中一部分的同位语，以追求语言表达得更加精炼。

Waves are able to bend around the edge of an obstacle in their path, a property called diffraction.

波能绕着其通道上的障碍物的边缘弯曲前进，这一特性称为绕射。

In every normal atom, the number of protons equals the number of electrons, a fact which is directly related to the electrical properties of the proton and the electron.

在每个正常的原子中，质子数等于电子数，这一点是与质子和电子的电性质直接相关的。

### 2.5.4 名词与介词常见的搭配模式

名词与介词常以“名词+of+A+介词+B”这种形式组合。其中 A 和 B 均为名词。掌握了这种模式而写出来的句子比较地道。

(1) 该名词来自普通的抽象名词。

This section deals with the advantages of the local roasting by microwave.

本节论述原位微波焙烧的优点。

The recovery rate is defined as the ratio of the metal produced and the the metal input.

回收率被定义为金属产出量和金属投入量之比。

(2) 该名词来自不及物动词。

The curve shows the variation of the leaching rate in the process with time.

这条曲线画出了浸出过程中浸出率随时间的变化情况。

The decrease rate of gold extraction with increased leaching temperature is b.

金提取率随浸出温度增加而下降的速度为 b。

(3) 该名词由及物动词演变而来。

These especial problems arise from the use of atomic energy as a source of power.

这些特殊问题是把原子能用作能源时产生的。

A comparison of a with b leads to the following equations.

把 a 和 b 比较，我们就得到了以下几个关系式。

## 2.6 代　词

在科技英语写作中，常会用物主代词做定语，与被修饰词之间存在主谓关系或动宾关系。

### 2.6.1 主谓关系

Its high solubility in water leads to significant reagent losses.

它的高溶解度导致了大量试剂的损失。

Physics is the most quantitative of the sciences, and we must become accustomed to its insistence upon accurate measurements and precise relationships if we are to appreciate its results.

物理学是各种科学中最讲究定量的，因此如果我们要理解其各种结果的话，我们必须习惯于它坚持要有精确的测量和各种精密的关系。

### 2.6.2 动宾关系

Many leaching technologies seem to neglect the performance of copper ion by ignoring their presence or avoiding much of the subject.

不少浸出技术似乎忽略了铜离子存在的行为，或是略去了许多内容。

The questions of environmental protection are not neglectable, and their study forms an important chapter in metallurgy.

环保问题不容忽视，关于它的研究在冶金学领域占据了重要的一章。

### 2.6.3 代词 one 的特殊用法

在科技英语写作中，需要表示人们、大家这一概念时，一般不用 people 表达，而是用被动语态或 one 做主语的主动语态。

No one can see the diffusion flux of copper ion.

大家不能看见铜离子的扩散流。

Before one studies a leaching system, it is necessary to define and discuss some important terms.

如果人们想要研究一个浸出系统，有必要先定义一些重要术语。

## 2.7 冠 词

在名词或名词短语前常有被称为限定词的词汇，这种用于限定名词范围的词汇就称为冠词。冠词与名词的单复数形式相关。例如：

a new result

many high-tech industry

the rising temperature

这些名词短语中的 a, the 等都是冠词。通常认为冠词“a”具有一个的含义，而冠词“the”具有“that”或“those”等词的含义，具有指示功能。

在实际写作时，冠词使用常见的错误有不加冠词或乱用冠词。通常冠词的使用可以遵循以下规律。

### 2.7.1 一般情况

（1）单数可数名词前需要加冠词：

A special type of rare earths resource

稀土资源的一种特别形式

An adequately oxidizing environment

一个充分氧化的环境

用 a 或是 an，取决于不定冠词后的第一个音素是不是元音。如果是元音，则需要用 an。

（2）在描述实验现象或结果时，如果已经在上文中描述过，而下文中需要再次描述时，需要使用冠词 the 加以描述。比较下面两句话所表述含义的差别：

A choice of extractants and aqueous solutions is influenced by both cost considerations and requirements of technical performance.

The choice of extractants and aqueous solutions is influenced by both cost considerations and requirements of technical performance.

除此之外，带有后置修饰语的特指的事物或表示一类人或物的单数可数名词前需要用定冠词。

（3）专有名词前通常不加冠词：

在科技英语写作中，涉及人名、地名、单位名称和国家名称时，通常单个词表示的国家名前不需要加冠词，而由三个或三个以上的普通单词构成的单位或国家名称前要加定冠词，如：

Northeastern University　东北大学

The Northwestern Polytechnical University 西北工业大学
The People's Republic of China 中华人民共和国

### 2.7.2 图示的说明文字可以不加冠词

Figure 2 Effect of temperature on the leaching rate
图 2 温度对浸出率的影响
该句子中省略了 effect 之前的 The。

### 2.7.3 不加冠词的情况

(1) 一些可数名词单数形式的泛指可以省略冠词：
He discovered the relationship between temperature and leaching rate.
他发现了温度与浸出率之间的关系。
(2) 人名的所有格之前一般不用冠词：
This equation is known as Newton's law.
这个公式被称为牛顿定律。

### 2.7.4 特殊情况

(1) 表示某个参数的单位时，常用定冠词：
The unit of temperature is the Kelvin.
温度的单位是开尔文。
(2) 几个名词并列时可以共用一个冠词：
The leaching rate depends on the particle size and shape.
浸出速率取决于颗粒大小和形状。
(3) 当表示了解一下、考察一下等抽象名词前一般使用不定冠词：
The prerequisite is a good knowledge of solvent extraction fundamentals.
先决条件是对萃取过程的基本内容有一个很好的了解。

## 2.8 常见词汇与词组用法

### 2.8.1 常见词汇和词组

(1) 引用数据（图片、表格）：

| | | | |
|---|---|---|---|
| show | present | list | report |
| illustrate | summarize | reveal | display |
| demonstrate | contain | describe | suggest |
| indicate | | | |

(2) 引用其他作者的结论/观点：

state
claim
demonstrate
note
argue
maintain
discuss
observe
suggest
assert
explain
expand
conclude
describe
publish
give
examine
presume
analyze
focus
hypothesize
develop
assume
contend
propose
find
recommend
based on…
show
report
in the view of…
according to…
use
study

（3）表达相似：

Similarly, …
Likewise, …`
X is similar/comparable to Y
In the same fashion, …
As in X, in Y…
X correspond to Y
Like X, Y…
The same…
X resemble Y

（4）表达对比/对照：

In contrast, …
Unlike X, Y…
Whereas/While…
In contrast to…
On the other hand, …
X is different from Y
…; however, …
…, but…
X differ from Y
X contrast with

（5）表达总结：

Overall, …
In general, …
On the whole, …
In the main, …
With… exception(s), …
The overall results indicate…
The results indicate, overall, that…

### 2.8.2 定义表示法

定义是对某事物的本质特征或一个概念的内涵和外延的确切而简要的说明。

2.8.2.1 种类

定义一般可以分为三种：

（1）个人定义。根据自己的立场、观点、经历等对某一事物或某一概念进行定义。

（2）广延定义。用各种方法进一步说明事物的本质、特点和概念的内涵和外延。这种定义有时很长，可以成为一篇文章。

（3）正式定义。这种定义通常只用一句话表示，其中包括了三个方面的内容，分别是所要下定义的事物或概念，它的类属以及它区别于同类中其他事物或概念的特点。这种定义的特点是能抓住事物的本质做出确切简要的说明，所以在科技文章中广泛采用这种定义。

#### 2.8.2.2 用于正式定义的句型

在科技写作中，用于正式定义的句型主要有以下三种：

（1）要说明的事物=类别词+表示该事物同其他类别词区分开的各种特性。具体为：

不定冠词+单数名词/不可数名词+be 动词+不定冠词+类别词+后置定语

这里一般有三类后置定语：定语从句、分词短语、介词短语（一般是 with 短语和 for 短语）。例如：

A force is a physical quantity that can affect the motion of an object.

力是能影响物体运动的一个物理量。

An thermocouple is a device which is used for raising temperature（或 used for raising temperature 或 for raising temperature）.

热电偶是提高温度的一种装置。

Stainless steel is an alloy which resists corrosion.

不锈钢是耐腐蚀的合金。

A factory is a place where products are manufactured.

工厂是制造产品的地方。

Iron is an element which has an atomic weight of 55.85（或 with an atomic weight of 55.85）.

铁是相对原子质量为 55.85 的一种元素。

需要注意的是，被定义的名词若为可数名词时，几乎总是用单数形式，而且在其面前一般只用不定冠词而不用定冠词。另外，定义短语中的主句是不带任何后置定语的，否则就不属于定义语句而成为说明句。如：

Efficiency is the ratio of the output energy of a machine to the input energy.（定义句）

The efficiency of a machine is the ratio of the output energy of that machine to the input energy.（说明句）

（2）A is/may be/can be defined as B：

Light metals are defined as metals with a relative density of less than 5.

轻金属被定义为相对密度小于 5 的金属。

Elasticity may be defined as the tendency of a body to return to its original state after being deformed.

弹性可以定义为物体形变后恢复其原状的趋势。

（3）By A is meant B 或 By A we mean B：

By recovery rate is meant the ability of a certain metallic element to recovery from ore.

所谓回收率指的是从矿石中回收某种金属元素的能力。

By frequency we mean the number of complete cycles per unit of time for a periodic quantity such as alternating current，sound waves，or vibrating objects.

所谓频率我们指的是诸如交流电、声波或一些振动物体这样的周期量每单位时间的完整周数。

### 2.8.3 名词所有格

在科技写作中，名词所有格的使用有以下五种情况需要了解并注意。

（1）当要表示科技术语缩写或某数值的复数形式时，需借助名词所有格的形式。例如：

emf's 或 EMF's

表示电动势的复数形式，但也可表示为 EMFs。

（2）用名词的所有格形式来表示动名词复合结构中动名词的逻辑主语；也可表示来自动词的一个名词的逻辑主语或宾语。例如：

We know of this rare earth's acting as an additive to some alloy.

我们知道这种稀土元素的作用像合金中的添加剂。

In this leaching system, the probability of copper ion's developing a local defect is very small.

在这个浸出体系中，铜离子产生原位缺陷的概率是很小的。

This is due in part to the cathode's rotation.

这部分原因是阴极的旋转。

The catalyzer's application in this field is becoming ever wider.

催化剂在这一领域中的应用越来越广泛。

（3）用所有格表示年代。例如：

in the 1990's 或 the 1990s

都可以表示 20 世纪 90 年代。

（4）在科技英语写作对非生命物而言，使用所有格的情况是比较少的，一般情况下使用"名词+of+名词"的形式代替。如：

the earth's surface（科普文中）

the surface of the earth（正式科技英语写作）

（5）注意以下三大规律：

1）若属于由某人发现的定律、定理、原理、效应等，则使用所有格形式表示。如：

| | |
|---|---|
| Euler's equation 欧拉方程 | Ohm's law 欧姆定律 |
| Boltzmann's constant 玻耳兹曼常数 | Moore's law 摩尔定律 |

2）若属于由某人发明的东西，则不用所有格而采用复合名词形式。如：

| | |
|---|---|
| a Diesel engine | 狄赛尔内燃机 |
| the Kelvin scale | 开尔文温标 |
| a Bursen burner | 本生灯 |
| a Wheatstone bridge | 惠斯顿电桥 |

需要注意的是，如果是由两个人或多人发现或发明的东西，则也要用复合名词形式，如：

| | |
|---|---|
| the Stefan–Boltzmann law | 斯蒂芬-玻耳兹曼定律 |
| the Joule–Thompson effect | 焦耳-汤普逊效应 |

3）若属于由某人发明的方法、技术、化学反应等，则上述两种表达形式均可采用，这就要求读者观察标准的使用习惯。如：

| | |
|---|---|
| Bollman's technique | 博尔曼方法 |
| the Bechamp technique | 比钱普方法 |

### 2.8.4 近似值表示方法

（1）使用某些形容词、副词或介词：

| | | | |
|---|---|---|---|
| approximate | approximately | under | nearly |
| almost | over | slightly | a little |

The cable connecting two leaching cells is approximately 5 centimeters thick.
连接两个浸出槽的电缆的直径约为 5 厘米。
That wire has a length of slightly over 3 meters.
那根导线长 3 米多一点。
The leaching column stands under 10 meters high.
这浸出柱竖着高度不到 10 米。
（2）使用被动语态：

| | | | |
|---|---|---|---|
| 某物+ | be+ | thought(据认为)<br>believed(据信)<br>estimated(据估计)<br>said(据说)<br>known(公认)<br>assumed(假定)<br>shown(表示为) | +动词不定式短语 |

That film surface is estimated to have a thickness of 3 centimeters.
那钝化膜表面估计有 3 厘米厚。
The ion activity is believed to be 2.
离子活度被认为是 2。

### 2.8.5 表示分类的常用句型

（1）用主动态“fall into”表示：
Forces for this separation process fall into two classes.
这个分离过程的驱动力分为两类。
（2）用被动态表示：

| | | | |
|---|---|---|---|
| can<br>may<br>might | +be+ | divided<br>classified<br>categorized<br>grouped | into/as/im |

Smelt may be classified as two major groups.

熔体可分为两大类。(注意这里的“大”，英语里应该用 major，main，general 和 board 表示)

Smelts of each kind can be grouped into many versions.

每种熔炼炉可以分成许多种型号。

(3) 用“there be”表示：

There are two kinds of heating equipment.

加热设备有两种。

(4) 用“be of”表示：

Surface measurements are of two kinds：those made of electronic quantities and those made by quantities such as pressure，temperature，or flow rate.

表面测量分两类：一类是对电子量所进行的测量；另一类是对诸如压力、温度或流速这些其他的量用电子方法进行的测量。

### 2.8.6 保持批判性

作为一个科研工作者，对自己所使用的资料应保持批判性。需要对读到的内容保持谨慎，而不是看已经出版就默认其为绝对正确。在科研工作中保持批判性不是意味着只接受正确的东西，而是需要你与作者一起找出问题所在。有时可能需要你指出其他作者的错误。下面介绍一些该部分常见的例句。

(1) 强调过去研究的不足之处：

Previous studies of X have not dealt with…

Researchers have not treated X in much detail.

Such expositions are unsatisfactory because they…

Most studies in the field of X have only focused on…

Half of the studies evaluated failed to specify whether…

The research to date has tended to focus on X rather than Y.

Most empirical studies of X have relied upon small sample sizes.

However, these studies used non-validated methods to measure…

The existing accounts fail to resolve the contradiction between X and Y.

Most studies of X have only been carried out in a small number of areas.

However, much of the research up to now has been descriptive in nature.

Small sample sizes have been a serious limitation for many earlier studies.

None of the studies reviewed appear to have controlled for the effects of...

The generalisability of much published research on this issue is problematic.

This general lack of methodological rigour may put in question the results of...

However, few writers have been able to draw on any structured research into...

The vast majority of researchers have not considered the interaction effects of...

There are obvious difficulties in accepting the reliability of self-report information.

However, these results were limited to X and are therefore not representative of...

Most of the research on the association between X and Y is flawed methodologically.

The experimental data are rather controversial, and there is no general agreement about...

Although extensive research has been carried out on X, no single study exists which adequately...

---

| | |
|---|---|
| Most studies of X | have only focused on... |
| | are unsatisfactory because they... |
| | fail to estimate economic rates of... |
| | have only investigated the impact of... |
| | have not included variables relating to... |
| | are limited by weak designs and a failure to address... |
| | have only been carried out in a small number of areas. |

---

（2）强调实证研究的不足：

The study suffers from...

The paper fails to specify...

No attempt has been made to...

The study makes no attempt to...

A major problem with this experiment was that...

No attempt was made to quantify the association between X and Y.

The scope of this research was relatively narrow, being primarily concerned with...

Smith's study of X is considered to be the most important, but it does suffer from the fact that...

However, these results were based upon data from over 30 years ago and it is unclear whether...

---

| | | |
|---|---|---|
| The paper | fails to | specify... |
| | does not | quantify... |
| | makes no attempt to | separate... |
| | | compare... |
| | | account for... |
| | | suggest why... |
| | | analyse how... |

---

| | | |
|---|---|---|
| The paper | fails to<br>does not<br>makes no attempt to | ascertain whether...<br>distinguish between...<br>explain the meaning of...<br>provide information on...<br>address the question of...<br>assess the effectiveness of...<br>use a standardised method of...<br>give sufficient consideration to...<br>consider the long term impact of...<br>offer an adequate explanation for...<br>engage with current discourses on...<br>determine the underlying causes of...<br>systematically review all the relevant literature. |

| | | | |
|---|---|---|---|
| (However,) | the study<br>the paper | suffers from | selection bias.<br>limited sample size.<br>poor external validity.<br>multiple design flaws.<br>an overemphasis on...<br>serious statistical flaws.<br>insufficient sample size.<br>inconsistent definitions.<br>poorly developed theory.<br>historical and cultural bias.<br>methodological limitations.<br>serious sampling problems.<br>a lack of clarity in defining...<br>inadequate research design.<br>considerable design limitations.<br>the use of poorly matched controls.<br>a paucity of standardised measures.<br>notable methodological weaknesses.<br>fundamental flaws in research design.<br>lack of a strong theoretical framework.<br>certain ambiguities at the conceptual level.<br>an over-reliance on self-report methodology.<br>a restricted range of methodological approaches.<br>shortcomings in the methods used to select cases. |

| | | |
|---|---|---|
| Smith<br>The study<br>The report | overlooks<br>fails to acknowledge<br>makes no attempt to consider | the impact of…<br>the reasons for…<br>the evidence for…<br>the contexts in which…<br>several key aspects of…<br>the variable nature of…<br>other explanations for…<br>the complex nature of…<br>the potential impact of…<br>the social dimension of…<br>the dynamic aspects of…<br>the underlying causes of…<br>the ethical implications of…<br>the important role played by…<br>the demographic factors that…<br>the broader implications of how…<br>the unique complexities faced by…<br>the contextual factors that influence… |

| | |
|---|---|
| No attempt has been made to | determine whether…<br>investigate whether…<br>estimate the risk of…<br>quantify the degree of…<br>model the dynamics of… |

| | |
|---|---|
| However | the analysis is largely superficial, based solely on…<br>the sample size in this study was relatively small…<br>this research has a number of methodological weaknesses.<br>the degree of X experienced by patients was not measured.<br>a major weakness with this study is that there no control for X.<br>a major problem with this experiment was that no control of X was used.<br>one of the problems with the instrument the researchers used to measure X was…<br>the main methodological weakness is that X was only monitored for 12 months. |

（3）说明理论或观点的局限性：

The main weakness with this theory is that...
The key problem with this explanation is that...
However, this theory does not fully explain why...
One criticism of much of the literature on X is that...
Critics question the ability of the X theory to provide...
However, there is an inconsistency with this argument.
There are limits to how far the concept of X can be taken.
A serious weakness with this argument, however, is that...
However, such explanations tend to overlook the fact that...
One question that needs to be asked, however, is whether...
One of the main difficulties with this line of reasoning is that...
Smith's argument relies too heavily on qualitative analysis of...
Smith's interpretation overlooks much of the historical research...
Many writers have challenged Smith's claim on the grounds that...
The X theory has been criticized for being based on weak evidence.
Smith's analysis does not take account of X, nor does he examine...
It seems that Jones' understanding of the X framework is questionable.
Aspects of X's theory have been criticized at a number of different levels.
The existing accounts fail to resolve the contradiction between X and Y.
One of the limitations with this explanation is that it does not explain why...
A final criticism of the theory of X is that it struggles to explain some aspects of...
The X theory has been vigorously challenged in recent years by a number of writers.
A second criticism of the hypothesis draws upon research evidence which suggests...
The X hypothesis has been questioned on the basis of some conflicting experimental findings.

| | |
|---|---|
| The theory is unable to | predict... |
| | explain why... |
| | fully account for... |
| | adequately explain the... |
| | explain what happens when... |
| | make any useful prediction about... |
| | explain the differences observed when... |
| | provide a comprehensive explanation for... |

| | |
|---|---|
| The current model of X suffers from | poor scalability. |
| | unnecessary complexity. |
| | lack of empirical support. |
| | several methodological problems. |
| | certain weaknesses that hinder its ability to... |

(4) 说明试验方法或试验操作的局限性：

One major drawback of this approach is that...

Selection bias is another potential concern because...

Perhaps the most serious disadvantage of this method is that...

The main limitation of biosynthetic incorporation, however, is...

Non-government agencies are also very critical of the new policies.

All the studies reviewed so far, however, suffer from the fact that...

Critics of laboratory-based experiments contend that such studies...

Another problem with this approach is that it fails to take X into account.

Difficulties arise, however, when an attempt is made to implement the policy.

Critics have also argued that not only do surveys provide an inaccurate measure of X, but the...

Nevertheless, the strategy has not escaped criticism from governments, agencies and academics.

Many analysts now argue that the strategy of X has not been successful. Jones(2003), for example, argues that...

| | |
|---|---|
| However, all the previously mentioned methods suffer from some serious | drawbacks.<br>limitations.<br>weaknesses.<br>shortcomings.<br>disadvantages. |

| | |
|---|---|
| However, | this method of analysis has a number of limitations.<br>this method does involve potential measurement error.<br>approaches of this kind carry with them various well known limitations.<br>questions have been raised about the reliability of self-report methods. |

| | |
|---|---|
| Selection bias is another(potential) | risk.<br>concern.<br>problem.<br>limitation.<br>weakness.<br>threat to internal validity.<br>limitation of systematic reviews. |

(5) 对其他作者的结论提出批评：

Smith fails to grasp that...

Smith's interpretation overlooks...

Smith overlooks a number of important sources.

Smith fails to acknowledge the social aspects of...

However, Smith's accounts are clearly ideological.

Although Smith has argued that... she neglects to note that...

Many aspects of Smith's interpretation have been questioned.

Smith's meta-analysis has been subjected to considerable criticism.

The most important of these criticisms is that Smith failed to note that...

The most convincing rebuttal of Smith's interpretations has been written by...

Smith's decision to priorities X as the primary cause of Y has been widely attacked.

The scope of this research was relatively narrow, being primarily concerned with...

Smith's study of X is considered to be the most important, but it does suffer from the fact that...

---

| | |
|---|---|
| (However,) | the paper does not address...<br>Smith fails to fully define what...<br>a major criticism of Smith's work is that...<br>Jones fails to acknowledge the significance of...<br>the author overlooks the fact that X contributes to Y.<br>what Smith fails to do is to draw a distinction between...<br>the main weakness of the study is the failure to address how...<br>Smith's paper would appear to be over ambitious in its claims.<br>another weakness is that we are given no explanation of how...<br>the research does not take into account pre-existing...such as...<br>the study fails to consider the differing categories of damage that...<br>the author offers no explanation for the distinction between X and Y.<br>Smith makes no attempt to differentiate between different types of X. |

---

| | | |
|---|---|---|
| Smith<br>The paper<br>The book | fails to<br>does not<br>makes no attempt to | specify...<br>quantify...<br>compare...<br>separate...<br>account for...<br>suggest why...<br>analyse how...<br>ascertain whether... |

| | | |
|---|---|---|
| | | distinguish between... |
| | | explain the meaning of... |
| | | provide information on... |
| | | address the question of... |
| Smith | fails to | assess the effectiveness of... |
| The paper | does not | use a standardised method of... |
| The book | makes no attempt to | give sufficient consideration to... |
| | | consider the long term impact of... |
| | | offer an adequate explanation for... |
| | | engage with current discourses on... |
| | | determine the underlying causes of... |
| | | systematically review all the relevant literature. |

| | | |
|---|---|---|
| | | the impact of... |
| | | the reasons for... |
| | | the evidence for... |
| | | the contexts in which... |
| | | several key aspects of... |
| | | the variable nature of... |
| | | other explanations for... |
| | | the complex nature of... |
| | overlooks | the potential impact of... |
| Smith | fails to acknowledge | the social dimension of... |
| | makes no attempt to consider | the dynamic aspects of... |
| | | the underlying causes of... |
| | | demographic factors that... |
| | | the ethical implications of... |
| | | the important role played by... |
| | | the broader implications of how... |
| | | the unique complexities faced by... |
| | | the contextual factors that influence... |

（6）提供建设性意见：

The study would have been more interesting if it had included...

These studies would have been more useful if they had focused on...

The study would have been more relevant if the researchers had asked...

The questionnaire would have been more useful if it had asked participants about...

The research would have been more relevant if a wider range of X had been explored.

| | | | | | |
|---|---|---|---|---|---|
| The study<br>The findings<br>Smith's paper<br>Her conclusions | would have been<br>might have been | more<br>much more<br>far more | useful<br>original<br>relevant<br>convincing<br>interesting | if he/she had<br>if the author had | used...<br>included...<br>adopted...<br>provided...<br>considered... |

A more comprehensive study would include all the groups of...

A better study would examine a large, randomly selected sample of societies with...

A much more systematic approach would identify how X interacts with other variables that...

(7) 评估其他作者工作的积极面：

This article provides a valuable insight into...

Overall, X's study is a powerful explanation of...

Smith's research is valuable to our understanding of...

The first major fieldwork project that was started in X was...

In his seminal text, Smith devoted some attention to....

One of the most influential accounts of X comes from Smith(1986)...

Smith's synthesis remains one of the most comprehensive studies of...

Smith makes an interesting contribution with regard to the impact of...

In a well-designed and robust study, Smith(1998) examined data from...

A good summary of the classification of X has been provided in the work of...

The pioneering work of Smith remains crucial to our wider understanding of...

The most comprehensive study of X during this period has been undertaken by...

Smith, in his comprehensive two-volume biography of X, devoted a substantial section to...

Smith's study is of great significance as it marks the first attempt to assess the broader impact of...

A more substantial approach to the longer-term significance of X can be found in Smith's recent article in...

| | | | |
|---|---|---|---|
| Smith(1990) | offers<br>provides<br>presents | a useful<br>a detailed<br>an original<br>an insightful<br>an extensive<br>an interesting<br>a comprehensive<br>a contemporary | analysis of... |

| | | | | |
|---|---|---|---|---|
| In his<br>In her<br>In this | useful<br>timely<br>detailed<br>thorough<br>excellent | study( of X) ,<br>survey( of X) ,<br>analysis( of X) ,<br>examination( of X) ,<br>investigation( into X) , | Smith( 2012)<br>Jones( 2014) | found…<br>concluded that…<br>was able to show… |

| | | | |
|---|---|---|---|
| Smith's | seminal<br>landmark<br>thoughtful<br>innovative<br>pioneering<br>influential<br>informative<br>fascinating<br>wide-ranging<br>comprehensive<br>ground-breaking | study<br>analysis | provides a valuable insight into…<br>makes a valuable contribution with regard to…<br>remains crucial to our wider understanding of…<br>is of great significance as it marks the first attempt to… |

| | | | | |
|---|---|---|---|---|
| In his<br>In her | seminal<br>thoughtful<br>innovative<br>pioneering<br>influential<br>informative<br>wide-ranging<br>comprehensive<br>ground-breaking | study | Smith | argues that…<br>provides a valuable insight into…<br>makes a valuable contribution with regard to… |

（8）引入其他作者的批评：

Jones( 2003) has also questioned why…

However, Jones( 2003) points out that…

The author challenges the widely held view that…

Smith( 1999) takes issue with the contention that…

The idea that… was first challenged by Smith( 1992).

Smith is critical of the tendency to compartmentalise X.

However, Smith( 1967) questioned this hypothesis and…

Smith( 1980) broke with tradition by raising the question of…

Jones( 2003) has challenged some of Smith's conclusions, arguing that…

Another major criticism of Smith's study, made by Jones(2003), is that...

Jones(2003) is critical of the conclusions that Smith draws from his findings.

An alternative interpretation of the origins of X can be found in Smith(1976).

Jones(2003) is probably the best-known critic of the X theory. He argues that...

In her discussion of X, Smith further criticises the ways in which some authors...

Smith's decision to reject the classical explanation of X merits some discussion...

In a recent article in Academic Journal, Smith(2014) questions the extent to which...

The latter point has been devastatingly critiqued by Jones(2003), who argues that...

A recently published article by Smith et al. (2011) casts doubt on Jones' assumption that...

Other authors (see Harbison, 2003; Kaplan, 2004) question the usefulness of such an approach.

Smith criticised Jones for his overly restrictive and selective definition of X which was limited to...

Smith's analysis has been criticised by a number of writers. Jones(1993), for example, points out...

| | |
|---|---|
| Smith | criticises...<br>questions...<br>challenges...<br>is critical of...<br>casts doubt on...<br>points out that...<br>takes issue with...<br>raises a number of questions about... |

### 2.8.7 分类和列表

在科技英语写作中，当我们需要对某事、物进行分类时，通常是根据这些事物所具有的某一共同性质为基准进行分类的。为了做这样的工作，我们需要理解它们作为一个类所共有的某些特性和特征。分类也是理解事物之间差异的一种方式。在写作中，分类是很常见的作为向读者介绍新话题的一种方式。随着定义的编写，函数分类可以用在文章的开头部分，也可以用在较长的文章中。而需要列表的情况，通常是我们需要系统性的表述一系列的内容，列表的顺序可以表示重要性的排序。

（1）常见的表示分类的语句：

X can be classified into Xi and Xii.

X can be categorised into Xi, Xii and Xiii.

There are two main types of X: Xi and Xii.

Different methods have been proposed to classify...

Generally, X provides two types of information: Xi and Xii.

It has become commonplace to distinguish 'Xi' from 'Xii' forms of X.

Bone is generally classified into two types: Xi bone, also known as…, and Xii bone or…

The theory distinguishes two different types of X, i. e. social X and semantic X(Al-Masry, 2013).

Smith's systematic treatises may be grouped into several divisions: logic, psychological works…

The works of Smith fall under three headings: (1) dialogues and…; (2) collections of facts; and (3) …

There are two basic approaches currently being adopted in research into X. One is the Y approach and the other is…

| | | |
|---|---|---|
| X may be divided into | three main | classes.<br>sub-groups.<br>categories. |

| | | | |
|---|---|---|---|
| X may be classified | in terms of<br>on the basis of<br>according to<br>depending on | Y | into Xi and Xii. |

(2) 具体的分类：

Smith draws a distinction between…

Smith's Taxonomy is a multi-tiered model of classifying X.

Smith(2006: 190) categorised X as either a) …, b) …, or c) …

Jones(1987) distinguishes between systems that are a) …, b) …, or c) …

A third method, proposed by Smith et al. (2010), bases the classification on a…

To better understand X, Smith et al. (2011) classified Y into three distinct types using…

In Jones' system, individuals were classified as belonging to upper or lower categories of…

Smith's Taxonomy is a classification system used to define and distinguish different levels of…

Smith and Jones(2003) argue that there are two broad categories of Y, which are: a) …and b) …

For Aristotle, motion is of four kinds: (1) motion which…; (2) motion which…; (3) motion which…; and(4) motion which…

| | |
|---|---|
| In the traditional system, X is graded | in terms of…<br>on the basis of…<br>according to whether… |

| | |
|---|---|
| Smith's taxonomy is | used to classify...<br>a hierarchical model for classifying...<br>a well-known description of levels of...<br>a classification of learning objectives...<br>a widely acknowledged classification system useful for...<br>a multi-tiered model of classifying X according to different levels of... |

| | | | |
|---|---|---|---|
| Smith(1966) | divided<br>classified<br>grouped | Xs | into two broad types: Xis and Xiis. |

| | | |
|---|---|---|
| Smith(1996) describes | four basic kinds of validity: | logical, content, criterion and construct. |

（3）对分类系统的评价（积极的或中性的）：

| | |
|---|---|
| This system of classification | includes...<br>allows for...<br>helps distinguish...<br>is useful because...<br>is very simple and...<br>provides a basis for...<br>has clinical relevance.<br>was agreed upon after...<br>can vary depending on...<br>is still respected and used.<br>is particularly well suited for...<br>has withstood the test of time.<br>is a convenient way to describe...<br>has been broadened to include...<br>was developed for the purpose of...<br>is more scientific since it is based on... |

（4）对分类系统的评价（消极的）：

| | |
|---|---|
| This system of classification | is misleading.<br>is now out of date.<br>can be problematic. |

| | |
|---|---|
| This system of classification | is in need of revision.<br>poses a problem for…<br>is not universally used.<br>is somewhat arbitrary.<br>is simplistic and arbitrary.<br>has relevance only within…<br>has now been largely abandoned.<br>is obsolete and tends to be avoided.<br>has limited utility with respect to… |

（5）说明列表的方法：

This topic can best be treated under three headings: X, Y and Z.

The key aspects of waster management can be listed as follows: X, Y and Z.

There are two types of effect which result when a process undergoes X. These are…

The Three Voices for Mass is divided into six sections. These are: the Kyrie, Gloria, …

There are three reasons why the English language has become so dominant. These are:

Appetitive stimuli have three separable basic functions. Firstly, they…Secondly, they…

This section has been included for several reasons: it is…; it illustrates…; and it describes…

The disadvantages of the new approach can be discussed under three headings, which are: …

During his tour of Britain, he visited the following industrial centres: Manchester, Leeds, and…

The Mass for Four Voices consists of five movements, which are: the Kyrie, Gloria, Credo, Sanctus and Agnus Dei.

(6)引用他人的列表：

Smith and Jones(1991) list X, Y and Z as the major causes of infant mortality.

Smith(2003) lists the main features of X as follows: it is A; it is B; and it has C.

Smith(2003) argues that there are two broad categories of Y, which are: a) …and b) …

Smith(2003) suggests three conditions for X . Firstly, X should be…Secondly, it needs to be…

For Aristotle, motion is of four kinds: (1) motion which…; (2) motion which…; (3) motion which…; and(4) motion which…

### 2.8.8 对比或比较

通过了解两件事情之间的相似点和不同点，可以增加我们对两者的理解并且学习更多的知识。而对比通常涉及一个分析的过程，在这个过程中我们比较具体的部分和整体。比较也可能是一个初级阶段评估。例如，通过比较A和B的具体方面，我们可以决定哪个更重要或有价值。许多具有对比或比较功能的段落，其开头一般都是一个概述性的句子。

（1）说明不同点：

X is different from Y in a number of respects.

X differs from Y in a number of important ways.

There are a number of important differences between X and Y.

Areas where significant differences have been found include X and Y.

In contrast to earlier findings, however, no evidence of X was detected.

A descriptive case study differs from an exploratory study in that it uses...

Jones(2013) found dramatic differences in the rate of decline of X between Y and Z.

Women and men differ not only in physical attributes but also in the way in which they...

The nervous systems of Xs are significantly different from those of Ys in several key respects.

| | | | |
|---|---|---|---|
| Smith(2003) | found<br>observed | only slight<br>minor<br>distinct<br>significant<br>notable<br>considerable<br>major | differences between X and Y. |

（2）说明相同点：

Both X and Y share a number of key features.

There are a number of similarities between X and Y.

The effects of X on human health are similar to those of Y.

Both X and Y generally take place in a "safe environment".

These results are similar to those reported by(Smith et al. 1999).

This definition is similar to that found in(Smith, 2001) who writes:

The return rate is similar to that of comparable studies(e. g. Smith et al. 1999).

The approach used in this investigation is similar to that used by other researchers.

Numerous studies have compared Xs in humans and animals and found that they are essentially identical.

| | | |
|---|---|---|
| The mode of processing used by the right brain | is similar to that<br>is comparable to that<br>is comparable in complexity to that | used by the left brain. |

（3）用比较级的形式在单句中进行比较：

In the trial, women made fewer errors than men.

Women tend to have greater/less verbal fluency than men.

Adolescents are more/less likely to be put to sleep by alcohol than adults.

Further, men are more/less accurate in tests of target-directed motor skills.

Women tend to perform better/worse than men on tests of perceptual speed.

Women are faster/slower than men at certain precision manual tasks, such as...

The part of the brain connecting the two hemispheres may be more/less extensive in women.

Women are more/less likely than men to suffer aphasia when the front part of the brain is damaged.

| | | |
|---|---|---|
| Women | are more/less likely to perform well<br>are more/less accurate in tests of X<br>make more/fewer errors in tests of X<br>may be more/less susceptible to X<br>tend to perform better/worse in tests of X<br>tend to have greater/less verbal fluency | than men. |

（4）在两个句子中说明不同点：

| | | |
|---|---|---|
| It is very difficult to get away from calendar time in literate societies. | By contrast,<br>In contrast,<br>On the other hand, | many people in oral communities have little idea of the calendar year of their birth. |

| | |
|---|---|
| According to some studies, X is represented as...( Smith, 2012; Davis, 2014) | Others propose...( Jones, 2014; Brown, 2015) |

| | |
|---|---|
| Smith( 2013) found that X accounted for 30% of Y. | Other researchers, however, who have looked at X, have found...Jones( 2010) , for example, ... |

| | |
|---|---|
| Zhao( 2002) reports that... | However, Smith's( 2010) study of Y found no... |

（5）在两个句子中说明相同点：

| | | |
|---|---|---|
| Young children learning their first language need simplified input. | Similarly,<br>Likewise,<br>In the same way, | low level adult L2 learners need graded input supplied in most cases by a teacher. |

| | | |
|---|---|---|
| Smith argues that...<br>Al-Masry( 2003) sees X as... | Similarly,<br>Likewise,<br>In the same vein, | Jones( 2013) asserts that...<br>Wang( 2012) holds the view that...<br>Smith( 1994) in his book XYZ notes... |

### 2.8.9 描述趋势的表示法

趋势是指事物随着时间发展或变化的总方向。通过趋势可以对未来变化进行预测。趋势和预测通常用横轴表示时间的线图来表示。以下是一些常用来描述趋势和预测的语句。

（1）描述趋势：

| | | | |
|---|---|---|---|
| The graph shows that there has been a<br>Figure 2 reveals that there has been a | slight<br>gradual<br>slow<br>steady<br>marked<br>steep<br>sharp | growth<br>increase<br>rise<br>decrease<br>fall<br>decline<br>drop | in the number of divorces in England and Wales since 1981. |

（2）描述图中的高点或低点：

Copper production peaked in 1985.

Gold production reached a( new) low in 1990.

The peak age for committing a crime is 18.

The number of live births outside marriage reached a peak during the Second World War.

（3）预测趋势：

| | | | |
|---|---|---|---|
| The rate of Z<br>The amount of Y<br>The number of Xs | is likely to<br>is projected to<br>is expected to<br>will probably | fall<br>grow<br>increase<br>level off<br>drop sharply<br>remain steady<br>decline steadily | after 2020. |

（4）强调图表中的数据：

| | | |
|---|---|---|
| What stands out in this | table<br>chart<br>figure | is the growth of…<br>is the high rate of…<br>is the dominance of…<br>is the rapid decrease in…<br>is the steady decline of…<br>is the general pattern of…<br>is the difference between… |

### 2.8.10　定量描述

对于非英语为母语的人来说，进行定量描述可能较为困难，因为其中有很多短语单词的组合，比如介词和代词，这些很容易混淆。下面给出的许多短语也包含近似词，如：nearly, over half, less than, just over 等。

（1）描述分数或百分比：

Nearly half of the respondents(48%) agreed that...

Approximately half of those surveyed did not comment on...

Of the 270 participants, nearly one-third did not agree about...

Less than a third of those who responded(32%) indicated that...

The number of first marriages in the United Kingdom fell by nearly two-fifths.

Of the 148 patients who completed the questionnaire, just over half indicated that...

70% of those who were interviewed indicated that...

The incidence of X has been estimated as 10% following...

Since 1981, England has experienced an 89% increase in crime.

The response rate was 60% at six months and 56% at 12 months.

Returned surveys from 34 radiologists yielded a 34% response rate.

He also noted that less than 10% of the articles included in his study cited...

With each year of advancing age, the probability of having X increased by 9.6%($p=0.006$).

The mean income of the bottom 20 percent of U. S. families declined from $10,716 in 1970 to...

X found that of 2,500 terminations, 58% were among young women aged 15−24, of whom 62%...

| | | | |
|---|---|---|---|
| Well over | | | |
| Many more than | | | |
| More than | half | | |
| Just over | a third | of those surveyed | agreed that... |
| Around | a quarter | of the respondents | indicated that... |
| Approximately | 70% | of those who responded | did not respond to this question. |
| Almost | 50% | | |
| Just under | 40% | | |
| Fewer than | | | |
| Well under | | | |

（2）描述平均值：

The average of 12 observations in the X, Y and Z is 19.2 mgs/m...

Roman slaves probably had a lower than average life expectancy.

This figure can be seen as the average life expectancy at various ages.

The proposed model suggests a steep decline in mean life expectancy...

The mean age of Xs with coronary atherosclerosis was 48. 3 ± 6. 3 years.

Mean estimated age at death was 38. 1 ± 12. 0 years( ranging from 10 to 60+years).

The mean income of the bottom 20 percent of U. S. families declined from $ 10, 716 in 1970 to...

The mean score for the two trials was subjected to multivariate analysis of variance to determine...

| | |
|---|---|
| Roman slaves probably had a | much lower than average life expectancy. |
| The Roman nobility probably had a | much higher than average life expectancy. |

(3) 描述范围:

The respondents had practised for an average of 15 years( range 6 to 35 years).

The participants were aged 19 to 25 and were from both rural and urban backgrounds.

Rates of decline ranged from 2. 71- 0. 08cm day( Table 11) with a mean of 0. 97cm day.

They calculated ranges of journal use from 10. 7%-36. 4% for the humanities, 25%-57% for...

The evidence shows that life expectancy from birth lies in the range of twenty to thirty years.

The mean income of the bottom 20 percent of U. S. families declined from $ 10, 716 to $ 9, 833.

Most estimates of X range from 200. 000 to 700. 000 and, in some cases, up to a million or more.

At between 575 and 590 metres depth, the sea floor is extremely flat, with an average slope of...

(4) 描述比例:

Singapore has the highest proportion of millionaire households.

The annual birth rate dropped from 44. 4 to 38. 6 per 1000 per annum.

East Anglia had the lowest proportion of lone parents at only 14 per cent.

The proportion of live births outside marriage reached one in ten in 1945.

The proportion of the population attending emergency departments was 65% higher in X than...

### 2. 8. 11 因果关系的表示法

大量的学术工作中包括对问题的理解和提出解决方案。在研究生阶段,特别是在应用领域,学生通过寻找问题来学习。事实上,可以认为大量的学术问题源自问题。然而,除非对问题进行充分分析,否则无法提出解决办法,而这涉及对原因的彻底了解。下面列出了一些解释因果关系的语句。

(1) 通过动词解释因果关系:

| | | |
|---|---|---|
| Lack of iron in the diet | may cause<br>can lead to<br>can result in<br>can give rise to | tiredness and fatigue. |

| | | |
|---|---|---|
| Scurvy is a disease | caused by<br>resulting from<br>stemming from | lack of vitamin C. |

| | | |
|---|---|---|
| Much of the instability in X | stems from<br>is caused by<br>is driven by<br>can be attributed to | the economic effects of the war. |

(2) 通过名词解释因果关系：

One reason why Xs have declined is that...

A consequence of vitamin A deficiency is blindness.

X can have profound health consequences for older people.

The most likely causes of X are poor diet and lack of exercise.

The causes of X have been the subject of intense debate within...

(3) 通过介词短语解释因果关系：

| | | |
|---|---|---|
| Around 200,000 people per year become deaf | owing to<br>because of<br>as a result of<br>as a consequence of | a lack of iodine. |

(4) 通过句子联接解释因果关系：

| | | |
|---|---|---|
| If undernourished children do survive to become adults, they have decreased learning ability. | Therefore,<br>Consequently,<br>Because of this,<br>As a result(of this), | when they grow up, it will probably be difficult for them to find work. |

(5) 通过副词解释因果关系：

| | | |
|---|---|---|
| Malnutrition leads to illness and a reduced ability to work in adulthood, | thus | perpetuating the poverty cycle. |
| The warm air rises above the surface of the sea, | thereby | creating an area of low pressure. |

（6）可能的暂时性因果关系：

X might be attributed to...

X may be associated with...

One reason for this difference may be...

There is some evidence that X may affect Y.

In the literature, X has been associated with Y.

It is not yet clear whether X is made worse by Y.

The findings indicate that regular exercise could improve...

A high consumption of X could be associated with infertility.

X in many cases may be associated with certain bacterial infections.

X appears to be linked to...

This suggests a weak link may exist between X and Y.

The use of X may be linked to behaviour problems in...

The human papilloma virus is linked to most cervical cancer.

| | | |
|---|---|---|
| X may | have | caused Y. |
| | | given rise to Y. |
| | | brought about Y. |
| | | been an important factor in Y. |
| | | contributed to the increase in Y. |
| | | been caused by an increase in Y. |
| | | played a vital role in bringing about Y. |

| | |
|---|---|
| X may have been | due to Y. |
| | caused by Y. |
| | attributed to Y. |
| | brought about by Y. |

### 2.8.12 举例说明

作者可以给出具体的例子作为证据来支持他们的论点或观点。例子也可以用来帮助读者或听众理解不熟悉或难懂的概念。由于这个原因，在教学中也常使用例子。学生可能会被要求在他们的学习中举例来证明他们已经理解了一个复杂的问题或概念。

（1）例子作为句子中的主要信息：

| | | |
|---|---|---|
| A/An | classic<br>useful<br>notable<br>important<br>prominent<br>well-known | example of X is… |

For example, the word "leaching" used to mean "dissolved species in water".

For example, Smith and Jones(2004) conducted a series of semi-structured interviews in…

Young people begin smoking for a variety of reasons. They may, for example, be influenced by…

Another example of what is meant by X is…

This is exemplified in the work undertaken by…

This distinction is further exemplified in studies using…

An example of this is the study carried out by Smith(2004) in which…

The effectiveness of the X technique has been exemplified in a report by Smith et al. (2010).

This is evident in the case of…

This is certainly true in the case of…

The evidence of X can be clearly seen in the case of…

In a similar case in America, Smith(1992) identified…

This can be seen in the case of the two London physics laboratories which…

X is a good illustration of…

X illustrates this point clearly.

This can be illustrated briefly by…

By way of illustration, Smith(2003) shows how the data for…

These experiments illustrate that X and Y have distinct functions in…

（2）例子作为句子中的补充信息：

The prices of resources, such as copper, iron ore, oil, coal and aluminum, have declined in real terms over the past 20 years.

Young people begin smoking for a variety of reasons, such as pressure from peers or the role model of parents.

Pavlov found that if some other stimulus, for example the ringing of a bell, preceded the food, the dog would start salivating.

In Paris, Gassendi kept in close contact with many other prominent scholars, such as Kepler, Galileo, Hobbes, and Descartes.

Many diseases can result at least in part from stress, including: arthritis, asthma, migraine, headaches and ulcers.

（3）举例作为论据：

This case has shown that…

This has been seen in the case of…

The case reported here illustrates the…

Overall, these cases support the view that…

This case study confirms the importance of…

The evidence presented thus far supports the idea that…

This case demonstrates the need for better strategies for…

As this case very clearly demonstrates, it is important that…

This case reveals the need for further investigation in patients with…

This case demonstrates how X used innovative marketing strategies in…

Recent cases reported by Smith et al. (2013) also support the hypothesis that…

In support of X, Y has been shown to induce Y in several cases(Smith et al., 2001).

### 2.8.13 一些注意事项

下面列出一些读者在撰写科技论文时可能出现的问题，供读者参考。

（1）科技写作中易忽视的字母该大写的场合：

1）图表、定理、章节、参考文献等带有顺序的号码，如：

| | | | |
|---|---|---|---|
| Fig. 1-5 | 图 1-5 | Problem 2 | 题 2 |
| Table 2 - 3 | 表 2 - 3 | Reference [6] | 参考文献［6］ |
| Theorem 4 | 定理 4 | Chapter 7 | 第七章 |
| Eq. (3.36) | 式（3.36） | Section 1-8 | 第 1-8 节 |

2）文章中提到带有学位、学衔等时，其称呼的首要字母要大写，如：

The authors would like to acknowledge the excellent review of the entire manuscript by Dr. Edward Nelson.

I am indebted to Professors Glen Goff, W. W. Peterson, M. A. Miller, Robert Carroll, Irving Reed, and Irwin Lebow for their suggestions.

（2）英汉表达不一致的地方：

1）主语不一致：

Something is wrong with this smelter.

这个熔炼炉有毛病。

2）主从句中代词使用位置不同：

Metals expand when heated.

金属受热就会膨胀。

If it were moving at a speed lower than that given by Eq. (2-4), the electron would fall to the nucleus.

若电子以低于式（2-4）所给出的速度运行的话，它就会掉到原子核上。

英语中往往在状语从句中使用代词，而汉语则在主句中使用代词。

3）比较对象表达上不一致：

The conductivity of copper is higher than that of aluminium.

铜的电导率比铝高。

（3）标点符号问题：

1）英语的句号是小圆点而汉语的句号是小圆圈；英语中没有顿号，应该用逗号表示；若在引号的内容之后有标点符号时（逗号或句号），美式写法一般将标点符号放在引号内，而英式写法一般将标点符号放在引号外。如：

Every measurement must have two parts, a number to answer the question "how many?" and a unit to answer the question "Of what?".

每个测量必须有两个部分，一个是回答"多少?"这一问题的数字，另一个是回答"有关什么?"这一问题的单位。

So as to facilitate better understanding of the smelting process, comments are appended to selected instruction lines by preceding them with a "; ".

为了有助于更好地理解熔炼工艺，给所选的指令线附加一些评述，其方法是在指令线前面放一个"；"。

The first two names in every directory are ". " and ". . ".

每个地址目录中的头两个名称是"．""．．"。

If the first character of the path name is "/", the starting directory is the root directory.

如果路径名称的第一个字符是"/"，那么起始目录就是基本目录。

One user might create file "example. c".

一位用户有可能创建文件"example. c"。

Only a file with ". com", ". exe", ". bat" extension can be executed.

只有带有".com""，exe"或".bat"扩展名的文件才能够被执行。

2）在缩略词"i. e."“viz.”和"e. g."后面应该加逗号。如：

The amount of solute you obtain in this way depends not only upon the volume of solution but also upon the concentration of solute, i. e. , the amount of solute in a given amount of solution.

以这种方法你所获得的溶质的量，不仅取决于溶液的体积，而且取决于溶质的浓度，也就是在给定量的溶液中溶质的量。

3）当有两个形容词修饰同一个名词时，一般在这两个形容词之间用逗号分开。如：

These techniques enable I/O devices to be treated in a standard, uniform way.

This protocol is suited for small, stable, sets of cooperating processes.

By examining a complete, real system, we can see how the various concepts discussed in this book relate both to one another and to practice.

In the end a clear, simple picture results.

4）在之前提到过，在"Figure X"说明的最后，不论是句子还是短语，均要加句号。在"Figure，Table，Theorem，Definition"等后面绝大多数人是不用黑点的，但要空两个空格：在"Table"的标题末结尾绝大多数人不用句号。

5）当句子以缩略词结尾时，因缩略词后已有黑点，所以句末不得再加句号。如：

误：The length of the cable is 5 mi.

正：The length of the cable is 5 mi.

这根电缆的长度为 5 英里。

（4）动词短语需要注意的情况：有些短语动词具有被动含义，所以不能再把它们表示成被动形式或用其过去分词作后置定语，主要有：

| | | | |
|---|---|---|---|
| consist of | 由……构成 | find use | 得到应用 |
| arise from | 由……引起 | act as | 用作为…… |
| result from | 由……引起 | function as | 用作为…… |
| serve as | 用作为…… | behave as | 用作为…… |

（5）人称代词的写法：通常而言，在涉及个人时，后面用人称代词时应该使用 he 或 she。

Even if a student can follow every line of every example in this book, that doesn't mean that he or she can solve problems unaided.

即使学生能看懂本书中每个例题的每一句话，这并不表明他就能独立解题了。

（6）单词的共用：当前后两个词语出现相同的单词且紧挨在一起时，该单词被共用。如：

A textbook author does not have available the material on which to base a reasonable conclusion as to whom the credit belongs.（to 是 as to 与 belongs to 共用）

教科书的作者不可能获得该感谢谁的合理结论所基于的材料。

In this case the transmission line behaves as if it were open-circuited.（as 是 behaves as 与 as if 共用）

在这种情况下，该传输线的作用好像是开路似的。

The model shows concentric orbits or, as they are now frequently referred to, rings or shells.（as 是 are referred to 与 as 共用）

该模型表示出了一些同心圆，也就是现在人们经常说的电子环或电子层。

（7）大小的关系：科技英语写作中，通常是小的在前，大的在后，但有时候大的在前，小的在后。如：

You may look up the correct symbol in Appendix B, Table 4.

你可以在附录 B 的表 4 中查出正确的符号。

See Ref. 9, pages 400 to 405, for further treatment.

关于进一步的论述，请见参考资料 9 的第 400 到 405 页。

（8）what 引导的名词从句：在由 "what" 引出的名词从句中，如果其主语比较长，则可以把连系动词 be 放在主语前。如：

In this way, we can identify what are the common elements of operating systems.

这样我们就能够确定操作系统的普通要素是什么了。

（9）"助动词或情态动词+be" 的写法：

What would be the magnitude and direction of the force acting on an electron placed at the point?

作用于放在该点的一个电子上的力的数值和方向是怎样的？

（10）数词的写法：如果一个句子以数字开头，则一般认为应该把数词拼写出来而不要用阿拉伯数字表示。现在国外出版社的一般做法是，一个数字在千位之后空一格而不用逗号隔开（如：24 000 000 而不用 24，000，000）；单位的商只能用一次斜杠，如果有三个单位，则要用负指数形式表示（如 $J \cdot mol^{-1} \cdot s^{-1}$）而不能写为（J/mol/s）。

（11）百分数的使用规则：

1）总数小于 25 的，不用百分数。

2）总数在 25 与 100 之间的，则百分数不用小数点。

3）总数在 100 与 100 000 之间的加一位小数点。

4）总数超过 100 000 的可以用两位小数点。

5）原数据应该包括进去，且百分数放在括号内。如：

Failure occurred in 301（8.1%）of the 3716 participants.

在 3716 位参赛者中有 301 位（占 8.1%）失败了。

（12）其他注意事项：

1）除了已被广泛接受的缩略词外，尽量少用缩略词。

2）两个段落之间空两个，段落下面有小标题时空三格；小标题后空两格；每段首行不必缩进去；每两行之间应该隔行打。

3）一个单词在一行末尾不得拆开，可以把该词全部放到下一行。

4）在“讨论”部分，一般应含有主要的信息、关键的评价、与其他研究的比较，从最主要的到次要的，以及结论。

5）有些读者写的句子往往是及物动词后没有宾语、该用被动形式的用了主动形式、主谓语不一致、名词单复数出错、冠词使用出错等现象。

6）除了祈使句外，一个句子一定要有主语。

7）标示作者的名字时，在各篇论文中应该一致，以便于他人查找、检索。

8）决不能同时一稿多投。

# 2.9 句子的表达方法

## 2.9.1 从句介绍

### 2.9.1.1 状语从句

（1）表示原因的连词，常见的有下面几个，其表示原因的语气由强至弱排列如下：

1）because（=in that）：它表示的原因构成了句子的最主要部分，从句往往放在主句后。

The active choride ion possesses more kinetic energy because it freely moves in the solution.

活泼的氯离子拥有更大的动力能量，因为它在溶液中自由运动。

2）since（=in that）：它表示的原因已为人们所知，或不如句子的其余部分重要；它比 as 更正式一些，从句一般放在主句前。

Since $k$ and $m$ are both constants, the ratio $k/m$ is constant.

既然 $k$ 和 $m$ 均为常数，$k/m$ 比值就是恒定。

Obviously, no current can flow between the anodes and the cathode since there is no circuit.

显然阴极和阳极之间不可能存在电流流动，因为根本没有电路。

The leaching kinetics of gold differs from silver leaching in that silver can form silver sulfide.

金的浸出动力学和银不同，因为银可以生成硫化银。

此句中需要注意的是，"in that" 只能处于主句之后，且还能译成"在于"。

3）as( =now that)：其用法与 since 类同，从句一般处在主语前。

As nickel ion can form complex with ammonia, it exhibits less activity than ferric ion.

由于镍离子能够和氨形成配合物，所以它比三价铁离子活度低。

Now that we have discussed the meaning of a graphical solution of a system of simultaneous equations and the method of plotting a line, we are in a position to find graphical solutions of systems of linear equations.

由于我们已讨论了联立方程组的图解的含义及画线的方法，我们现在能求出线性方程组的图解了。

4）for：由它引出的句子使人觉得所述的理由只是一种补充说明而已，它是一个并列连词，引出的句子为并列句而不是状语从句，它只能放在前一句的后面，偶尔也有单独成为一句的。

In previous chapters we did not use same kinetic functions to describe the leaching behavior of copper, for the derivative of each of these is a special form.

在前面几章我们并没有使用同样的动力学方程去描述铜的浸出行为，因为其各自的导数均为一种特殊的形式。

（2）表示"当…… 时候"的连词，以下三个词在科技文中使用时常见有如下的基本区别：

1）when：它引出的从句表示的动作往往为非延续性动作，而这时的主语的动作也多为非延续性动作。

When this happens, the $Cu_2O$ dissolves in the slag generated during copper making.

当这种情况发生时，在铜生产过程中 $Cu_2O$ 溶解于产生的渣中。

2）while：它引出的从句往往采用进行时态。从句中的动作一般持续一段时间，而主句的动作则多表示一点或一种状态。

It is possible for the copper to remain in the solution while it is dissolving.

当铜溶解时，它们能够留在溶液中。

While the switch is off, the redox potential increases.

当开关关闭时，氧化还原电位升高。

3）as：它引导的从句中的动作与主句的动词同时发生，并持续一段时间。

As acidic slags forms, the dissolution of other acidic oxides increase.

当酸性渣形成时，其他氧化物的溶解度提高。

（3）表示"虽然"的连词。在科技英语写作中常用的有以下几个：

1）although：一般用于正式的场合，同时可用于各种文体。

The foregoing provides a basis for this theorem, although it cannot be considered as a proof.

前面所讲的内容为这个定理提供了基础，虽然不能当作一种证明。

2）though：它一般用于非正式的口语或书面语中，该从句可采用特殊词序（在科技英语写作中一般是作表语的形容词放在 though 之前）。

Important though this kinetic law is, it is seldom used in practice.

该动力学定律虽然重要，但在实际中很少使用。

3）as：用于正式的文体中，从句一定要采用特殊词序（在科技英语写作中主要是作表语的作表语的形容词放在 as 之前）。

Active as the choride ions are, they play an important role in the formation of metal complexes.

氯离子很活泼，但它们在金属配合物生成方面非常重要。

4）while：它引导的从句侧重于对比；当主句和从句的句型相同时，一般把 while 译成"而"（若这是 while 从句放在主句前，则把"而"字译在主句之前）。

While $x$ can only lie between $-1$ and $+1$, there are an infinite number of values of $y$ for every value of $x$.

虽然 $x$ 只能处于 $-1$ 和 $+1$ 之间，但对应于每个 $x$ 值却有无限个 $y$ 值。

Blast-furnace copper smelting is given a rather brief treatment because it is a dying process while newer techniques such a continuous copper - making and solvent extraction are given extensive coverage

高炉熔炼铜由于是一种染色工艺，因此处理时间很短，而诸如连续制铜和溶剂萃取等较新的技术得到了广泛的应用。

At this time the kinetic energy approaches infinity, while the potential energy approaches the minimum.

这时动能趋于无穷大，而势能则趋于最小值。

Input A goes low while input B remains high

输入 A 变成低电位而输入 B 保持高电位。

While energy is the capacity to do work, power is the quantity of work in unit time.

能量是做功的能力，而功率则是单位时间内所做功的量。

（4）由 since, as, when, while, after, before 等引导的时间状语从句充当修饰名词的情况：

Let us consider the case when the torgue is zero.

让我们来考虑一下当转矩为零的情况。

The slag-forming stage is finished when the Fe in the matte has been lowered to about 1%.

当冰铜中的铁含量降低到 1%左右时，就完成了造渣阶段。

In the ten years since this book was first published, significant changes have been seen in metal-working.

自从本书首次出版以来的这十年间，在金属加工方面发生了巨大的变化。

2.9.1.2 同位语从句

（1）两个常用的句型：

1）毫无疑问：

there is no doubt/question that…

2）有证据表明：

there is evidence that…

（2）动宾译法的句型：

There is a growing awareness that these techniques are also of value in some other areas.

现在人们越来越认识到，这些技术在其他一些领域中也是很有价值的。

One of the main achievements is the recognition that properites of a material should be included in the analytic model.

主要的成就之一是人们认识到了材料的性质应包括在分析模型中。

（3）由名词从句转变成的同位语从句：

The question whether there is impurities in that lixivium will be discussed.

在浸出液中是否由杂质这个问题有待讨论。

The leaching rate has no guarantee how long this kind of efficiency will be worked.

浸出率得不到保证，这种效果能持续多久。

2.9.1.3 名词从句

在名词从句方面应重点掌握以下两大方面的一些句型：

（1）采用形式主语 it 的一些句型：

1）it+连系动词+表语+主语从句；

2）it+被动语态+主语从句；

3）it+不及物动词+主语从句。

（2）“what”从句的句型：

1）表示“什么、多大、哪个……”（即疑问代词的词义）：

We must understand what is meant by the slope of a function.

我们必须懂得函数斜率的含义。

Now we can determine what the intensity of the electric field is at that point.

现在我们能确定在该点电场强度有多大。

2）表示“……的”：

What this paper describes is worth reading.

这篇论文所述的内容值得一读。

A computer can do what it has been told to do.

计算机能做人们告诉它要做的事。

What we need is a thermocouple.

我们所需要的是一个热电偶。

Energy is what brings changes to materials.

能量是引起物质变化的东西。

This is close to what has been observed.

这接近于观察到的情况。

This illustrates what happens when a mixture of FeO, FeS and $SiO_2$ is heated to 1200℃.

这说明了 FeO、FeS 和 $SiO_2$ 的混合物加热到 1200℃时会发生什么。

It must be clear from what we have already learnt that men are much more intelligent than a machine.

从我们已了解的知识清楚认识到人要比机器高明得多。

3）表示“所谓的；通常所述的”。以被动语态为例，这一类的句型通常为：

What is called/termed/named/described as/known as/referred to as/spoken of as…

What we call a machine is really a kind of tool that can do work for man.

我们所称的机器实际上是能为人类做事的一种工具。

Fig. 1 shows what happens as a result.

图 1 显示了结果。

Leaching rate changes with what is described as overpotential.

浸出率随所谓的过电位而变化。

In 1895, a German physicist discovered what are now known as X rays.

在 1895 年，一位德国物理学家发现了现在所说的 X 射线。

Energy can transmit molecule from high concentration region to low concentration region by means of what is called diffusion.

利用所谓的扩散可把分子从高浓度区域传送到低浓度区域。

4）表示“现在/原来的样子”：

These steel factories are quite different from what they were.

这些钢厂与原来大不一样了。

In this case, the leaching rate is 3 times what it was.

在这种情况下，浸出率为原来的 3 倍。

#### 2.9.1.4 定语从句

在科技英语写作中应掌握的要点如下：

（1）最基本的定语从句（即关系代词在从句中作主语、宾语和定语以及关系副词在从句中作状语的情况）：

The instrument that can be used to measure current, voltage and resistance is called a multimeter.

能用来测量电流、电压、电阻的仪表叫做万用表。

The meter that we use to measure voltage is known as a voltmeter.

我们用来测量电压的仪表叫伏特表。

A target is the object whose position is to be determined.

目标就是其位置要加以确定的物体。

We must understand the reason why the thermocouple is used here.

我们必须懂得在这里使用热电偶的理由。

Ohm's law can be written as $V=IR$ where $V$ represents voltage, $I$ current and $R$ resistance.

欧姆定律可写成 $V=IR$，式中 $V$ 表示电压，$I$ 表示电流，$R$ 表示电阻。

（2）关系代词在从句中作介词宾语，而“介词+which”在从句中作状语的句型，侧重点放在如何正确选择介词方面，选择介词应从三个方面加以考虑：

1）与被限定名词的搭配要求；

2）从句中动词、名词或形容词所需的搭配要求；

3）由整个句子所要表达的概念来确定。

The rate at which feed enters the concentrate burner is measured by supporting the feed bins on load cells.（at 是与 rate 搭配使用的）

进料进入浓缩燃烧器的速度是通过在测压元件上支撑进料箱来测量的。

In a later chapter we will see the reasons for which different sources emit different types of spectrum.

在后面一章我们将看到不同的光源发射出不同类型光谱的原因。

It is these forms for which the most advanced reprocessing technology exists.

最先进的后处理技术就是以这些形式存在的。

A central pipe through which concentrate falls into the reaction shaft.

精矿通过中心管进入反应井。

The two elements of which water consists are the gases oxygen and hydrogen.

构成水的两种元素是气体氧和氢。

The rates at which oxygen and air flow into the concentrate burner are important flash furnace control parameters.

氧气和空气进入浓缩燃烧器的速度是重要的闪速炉控制参数。

It is controlled by adjusting the rate at which flux is fed to the solids feed dryer.

它是通过调节助熔剂被送入固体干燥机的速率来控制的。

The material of which this machine is made is iron.

制成这台机器的物质是铁。

Iron is one of the metals with which we are most familiar.（with 是词组 familiar with 所要求的）

铁是我们所熟悉的金属之一。

The substance in which there are many free electrons is called a conductor.（in 是根据整个句子的概念要求而确定的）

存在许多自由电子的物体称为导体。

（3）先行词为不定代词，或被序数词、形容词最高级或形容词 only，no，very 等修饰时不能用 which 而只能用 that（不过在 something 后也可以用 which）：

All that one need do is push the button.

人们只需按一下按钮。

Computers are the most efficient assistants that man has ever had.

计算机是人类所曾有的最有效的助手。

The first thing that will be done is to measure the lixivium of temperature.

要做的第一件事是测出浸出液的温度。

This is the only measure that we can take.

这是我们所能采取的唯一措施。

Anything that is hot radiates heat.

任何热的东西均辐射热量。

Close contact is all that is required.

只需要接触紧一点。

In this way the inserted meter will not affect the very thing that we wish to measure.

这样，接入的仪表并不会影响我们想要测量的参数。

(4) 在科技英语写作中关系词可省略的三种情况：

1) 关系代词在从句中作及物动词的宾语时（该词不一定是谓语）：

There are the kinds of grammatical structures and sentence patterns professional metallurgist prefer.

这些是冶金专家们喜欢采用的那种语法结构和句型。

The amount of gravitational pull a body produces depends on the amount of material in it.

一个物体产生的万有引力的大小，取决于它所含物质的多少。

This component carries out the instruction we give it.

这个部件执行我们给它发出的指令。

Fever usually is preceded by a chill such measures as blankets and hot water bottles seem unable to relieve.

发烧之前通常伴有用毛毯和热水袋这样的东西均无法减轻的发冷过程。

In this case the power in the load will be the maximum the source is capable of supplying.

在这种情况下，负载上的功率是电源所能提供的最大功率。

Each CPU has a very elementary set of functions it knows how to perform.

每个中央处理装置具有一套它懂得如何执行的极为基本的功能。

2) 在 way, distance, direction, reason, time, number of times, amount 等后可以省去关系副词或"介词+which"，这时也可用关系副词 that 引导从句：

Work is the product of the force and the distance a body moves.

功等于力与物体运动距离的乘积。

This is the main reason reverberatory furnaces continue to be shut down.

这是反射炉继续关闭的主要原因。

These diagram shows the way the leaching rate changes with time.

这个图说明了浸出率随时间变化的情况。

Reflections come back only from objects in the direction the antenna is pointed.

回波只能从处于天线所指的方向上的物体反射回来。

We may interpret the absolute as being the number of units a given number is from the origin, regardless of direction.

我们可以把绝对值解释为一个给定的数离原点的单位数，与方向无关。

If the slope of a straight line is 2, for every 1 unit we move to the right we have to move 2 units up, to stay on the graph.

若某直线的斜率为 2，则往右每移动一个单位，我们就得向上移动 2 个单位才能保持在图线上。

3) 当关系代词 which 在从句中作单个介词（而不是介词短语）的介词宾语并且此介词短语在从句中作状语时可以省去，其条件是一定要把介词置于从句末尾（这种省略情况在正式科技英语文章中少一些，而主要出现在科普英语文章中）：

Power is the rate work is done at.

功率是做功的速率。

The atomic mass of a metel element also depends on the proton number and neutron number.

金属元素的相对原子质量取决于其质子数和中子数。

Iron is one of the metals we are most familiar with.

铁是我们最熟悉的金属之一。

Like the sun, water is one of the necessities plants cannot go without.

像太阳一样，水是植物所离不开的必需品之一。

A battery or other source supplies a potential difference for the circuit it is connected to.

电池或别的电源为与其连接的电路提供一个电位差。

Liquids readily assume the shape of any vessel they are poured into.

液体容易呈现存放它们容器的形状。

We charge one pith ball with a rubber rod and another with the fur the rod was stroked with.

我们用一橡胶棒使一木髓球带电，而用该橡胶棒摩擦过的毛皮使另一个木髓球带电。

（5）which 引导修饰整个主句的非限定性定语从句的三种常见情况：

1）which 在从句中作主语：

The motion of a rocket does not depend on the presence of air, which was proved in 1916.

火箭的运行并不取决于空气存在，这在 1916 年就得到了证明。

Some ceramic materials are liable to absorb oxygen, which will adversely affect their insulating properties.

某些陶瓷材料容易吸收氧气，这会降低它们的绝缘性能。

A turned amplifier rejects signals far from the resonant frequency, which is often a considerable advantage.

调谐放大器可以抑制远离谐振频率的信号，这往往是一大优点。

The input may be connected to signal sources that have neither terminal grounded, which often proves to be convenient.

可以把输入连接到两个端点均不接地的信号源上，这往往证明是可行的。

2）which 在从句中作介词宾语的定语（which 等效于 this 或 that）：

This diffusion current continues until equilibrium is reached, at which time an internal barrier potential is built up.

这种扩散电流持续到平衡为止，此时内部的势垒电位就建立起来了。

This transfer continues until a uniform temperature is reached, at which point no further energy transfer is possible.

这一传递过程一直延续到温度均匀为止，那时能量便不可能进一步传递了。

The resistivity of most insulators decreases with an increase in temperature, for which reason insulated conductors must be kept at low temperature.

大多数绝缘体的电阻率随温度的上升而下降，所以绝缘了的导体必须处于低温状态。

3）which 在从句中作介词宾语（which 译为 this）：

We equate the mole ratio to the atom ratio, from which the simplest formula follows directly.

我们使摩尔比与原子比相等，由此直接得出了最简分子式。

The inactive material is $SO_2$, which acts as a support for the active components.

$SO_2$ 是非活性物质，可作为活性组分的支撑。

Some materials have a very high resistance, because of which they can be used as insulators to prevent the leakage of current.

有些物质的电阻率很高，因此可以用作绝缘体来防止漏电。

(6) 由 as 引导的定语从句：

1) as 引导的修饰整个主句的非限定性定语从句：

As we shall see, acids and bases play an important role in the functioning of the leaching system.

我们将会看到，酸和碱在浸出系统中起着十分重要的作用。

As we mentioned earlier, material can be used to solve systems of equations.

我们在早些时候提到过，矩阵可用来解方程组。

As these examples illustrate, ratios may compare quantities of the same kind.

正如这些例子所示，比率可以比较同种量。

Catalyst is manufactured by mixing together the active components and substrate to form a paste which is extruded and baked at-530℃ into solid cylindrical pellets or rings.

催化剂是将活性成分和底物混合在一起形成糊状物，在-530℃下挤压和烘烤成圆柱形的实心球团或环状物。

This equation is of great help in solving problems on acceleration, as will be shown later.

以后会证明，这个式子对求解有关加速度的题目是很有帮助的。

2) as 引导的修饰某个名词的限定性定语从句（往往与 such 和 the same 连用）：

This force produces the same effect as is produced by the simultaneous action of the given forces.

这个力产生的效应与给定的几个方向同时作用所产生的效应相同。

Such meters as we use to measure temperature are called thermometer.

我们用来测量温度的这类仪表称为温度计。

Such light nonlinearities as are found in vacuum tubes may be neglected in the small-signal case.

像真空管所呈现的这种微弱的非线性，在小信号情况下可以忽略不计。

3) 一种特殊结构形式为：as+过去分词/介词短语/副词。如：

As point out in the last chapter, the emf of a battery is generated by the chemical action within it.

正如上一章所指出的，电池的电动势是由其内部的化学作用产生的。

As shown in the last book, anodic protection of the coolers is required to minimize corrosion by the hot sulfuric acid.

如上一本书所指出的，我们需要对冷却器进行阳极保护，以减少热硫酸的腐蚀。

These leaching data are plotted on log-log paper, as shown in Fig. 2-6.

把这些浸出数据如图 2-6 所示那样画在双对数纸上。

The second law of motion as stated above is interesting but not especially useful.

上面所述的第二运动定律是很有趣的，不过并不特别有用。

The basic organization of such leaching system is as shown in Fig. 6.

这种浸出系统的基本组成如图 6 所示。

（7）由 that 引导的定语从句：

This kind of leaching reactor creates more fuel than it consumes.

这种浸出反应器产生的燃料比它消耗的多。

A system which provides more energy at the output than is given at the input is said to be active.

在输出端提供的能量比在输入端获得的能量大的系统就称为有源系统。

## 2.9.2 虚拟语气

（1）条件式的情况：

1）涉及现在或将来的情况：从句用一般过去时，主句用过去将来时。如：

If this copper were cast, the S and O would form $SO_2$ bubbles or blisters which give blister copper its name.

如果这个铜被铸造，S 和 O 将形成 $SO_2$ 泡沫或水泡，即为水泡铜。

2）涉及过去的情况：从句用过去完成时，主句用过去将来完成时。如：

If this method had been adopted(at that time), much time would have been saved.

如果当时采用了这种方法，就会节省很多时间。

3）涉及将来的情况：从句用“should+动词原形”或“were+动词不定词”，主句用一般将来时和过去将来时，有时也可用祈使句。如：

Should the nucleus have too few neutrons, the reverse reaction may/might take place.

要是原子核具有的中子数太少，就可能发生逆反应。

需要注意的是，本句中从句引导词 if 省去了，所以把 should 放在从句主语前，这一形式在科技写作中比较常见。

（2）出现在某些主语从句、宾语从句、表语从句和同位语从句中的情况。这种情况与主句中的某个及物动词、形容词或名词有关，所以一定牢记这些词。

常见的动词：demand, desire, insist, necessitate, order, propose, recommend, request, require, suggest。

常见的形容词：better, desirable, essential, imperative, important, impossible, natural, necessary, possible, reasonable。

常见的名词：condition, constraint, demand, recommendation, requirement, restriction, suggestion。

使用规则：这种虚拟语气的形式是从句谓语采用“（should+）动词原形”。

1）在主语从句中：

It is very important that all components(should) be checked in the initial stage of leaching.

极为重要的是所有的组分都要在浸出初始阶段中检验一下。

It is not necessary that the temperature be suddenly raised or declined.

没有必要突然提高或降低温度。

It is better that an analyst scrap his fine analysis, rather than he later see the mechanism scrapped.

最好是一位分析工作者宁可废弃他精密的分析，也不要在后来看到制造出来的装置被废掉。

It is not intended that this book stand alone as a course text.

编者并不打算把这一本书单独作为课程教科书。

It is suggested that this flowsheet design be modified at once.

我们建议应立刻修改这项流程设计。

It is my philosophy that text includes many examples, and that these examples be worked in sufficient detail so that the reader can follow each example from beginning to end.

我的看法是一本教科书应包括许多例子，并且这些例子应详细地推演出来，使得读者能从头至尾地看懂每个例子。

2）在宾语从句中：

This requires that the leaching temperature should be raised.

这里要求浸出温度必须提高。

Minkowski proposed that time or duration be considered as the fourth dimension supplementing the three spatial dimensions.

敏考斯基建议把时间或持续时间看成第四维以作为对立体三维的补充。

In the year 1791, the French Academy of Science suggested that the unit of length be based on the size of the earth.

法国科学院在1791年建议，长度的单位应以地球的大小为基础。

We recommend that he not try to absorb this chapter completely before preceeding to the subsequent chapters.

我们建议读者不必等完全掌握了这一章的内容后才去学习后面的章节。

This increase in resolution necessitates that more numerical information be obtained during the conversion process.

分辨能力的提高，要求在转换过程中必须获得更多的数值信息。

They consider it important that proper attention be paid to the algebraic signs of distances, velocities, and accelerations.

他们认为，重要的是应当注意距离、速度和加速度的代数符号。

The great power and versatility of catalyst and, consequently, their widespread application make it imperative that metallurgy engineering student obtain a working familiarity with that.

催化剂有着巨大威力和多种用途而且被广泛应用，因而冶金工程的学生必须熟悉催化剂的实用知识。

3）在表语从句中：

Our demand is that another leaching experiment(should) be made.

我们的要求是再做一个浸出实验。

Their suggestion is that this point be grounded.

他们的建议是，这个点应该接地。

It is clear that a necessary condition for the function $y$ which makes the integral a minium(or a maximum) is that it be a solution of Euler's equation.

显然，函数 $y$ 使该积分为最小值（或最大值）的必要条件是它必须是尤拉方程的一个解。

4）在同位语从句中：

The requirement that energy(should) be conserved must be fulfilled.

能量守恒这一要求一定要得到满足。

（3）出现在少数状语从句中的情况：

1）在 as if（as though）引导的从句中：当涉及现在（或将来）的情况时，谓语动词用一般过去时（be 用 were）；当涉及过去的情况时，谓语动词用过去完成时。如：

The wire carrying an electric current produces a magnetic field as if it were a magnet.

载流的导线产生一个磁场，好像是一块磁铁似的。

It appeared as if that leaching tanks had been damaged.

看起来浸出槽好像被损坏了。

2）在 whether（常省去）引导的从句中：最常见的句型是省去 whether 后动词 be 用原形倒放在从句主语之前，形成"be it…；be they…"这样的结构。如：

Any matter, be it air, water or wood, has weight.

任何物质，不论是空气、水，还是木头，均具有质量。

All these problems, be they easy or difficult, can be solved by this method.

所有这些题目，不论容易还是困难，都可以用这种方法来解。

3）在 lest 引导的从句中：在 lest 引导的从句中，从句谓语用"（should+）动词原形"。如：

Batteries should be kept in dry place lest electricity leak away.

应当把电池存放在干燥的地方以免跑电。

### 2.9.3 句子成分的强调

（1）采取强调句型"it is/was… that/which/who…"。这种巨型一般用来强调主语、宾语、状语和状语从句，当被强调的句子属于现在（或将来）的任一时态时，本句型应当使用 is 或 will be。当属于过去的任一时态时，该句型应该使用 was；另外，that 可用于强调任意情况，which 只能用来强调做主语或宾语的事物，而 who 只能用来强调做主语的人。这个句型一般可翻译成正是，是；但当强调引导名词从句或特殊问句的 what，when，why，how，where，which 等词时，则应翻译成到底，究竟。如：

It is the net force on an object that/which causes acceleration.

正是在物体上的净力引起了加速度。

It is the losses caused by diffusion that/which we must try to overcome.

我们必须尽力克服扩散引起的各种损耗。

It is when an object is heated that the average speed of molecules is increased.

正是当物体受热时，分子的平均速度提高了。

(2) 采用助动词 do/does/did。在这种句型一般用来强调一般过去时和一般现在时的谓语动词，译为的确，确实，一定。如：

These methods do work.

这些方法确实可行。

The moon does have gravity.

月球的确具有引力。

However, methods to decrease slag mass may do more harm than good.

然而，降低矿渣质量的方法可能弊大于利。

(3) 采用形容词 very。这种句型用来强调某个名词，译为就，最。如：

Catalyst is the very component that makes leaching process possible.

催化剂使得浸出过程成为了可能。

The leaching starts at the very moment we raise the temperature.

浸出过程在升温的那一刻就开始了。

In this way, when the sulfuric acid is added it does not change the very thing we wish to aquire.

这样，当加入硫酸时，它不会改变我们想要的那个结果。

(4) 采用倒装句：

1) 全倒装：

Moving round the nucleus are the negatively charged particles called electrons.

绕原子核旋转的是称为电子的带负电的微粒。

Based on this relationship between magnetism and electricity are motors and generators.

以磁和电之间的这一关系为基础的有电动机和发电机。

2) 部分倒装：

This driving force we call chemical energy.

这种驱动力我们称之为化学能。

Metallurgy makes possible a great many things.

冶金学使得许许多多的东西成为可能了。

### 2.9.4 句子成分的倒装

#### 2.9.4.1 完全倒装

(1) “表语+连系动词+主语”的句型：

Of wider application is the fact that matter usually expands when its temperature is increased and contracts when its temperature is decreased.

应用更为广泛的是物质通常的热胀冷缩现象。

Most interesting has been the fact that certain alloys become superconductors at rather high temperatures.

极为有趣的是某些合金在颇高的温度时能变成超导体。

Among the most noteworthy achievements at that time was the realization that light consists of electromagnetic waves.

当时最引人注目的成就之一是人们认识到了光是由电磁波构成的。

Basic to understanding the binary number system is a familiarity with power of 2.

理解二进制数系的关键是要熟悉 2 的各次幂。

(2)"状语+不及物动词+主语"的句型:

In the electrowinng cell occur such magic behavior as metal co-precipitation.

在电沉积槽里发生着像共同沉积的神奇行为。

The center of mass is the point through which passes the resultant of the reaction forces when a body is accelerated.

质量中心就是当物体被加速时各反作用力的合力所通过的那一点。

Out of Newton's studies in the analysis of the spectrum has come the whole technique of modern spectrum analysis which is the basis of research in present-day astronomy.

由于牛顿在频谱分析方面的研究,产生了现代频谱分析的整套技术,这是现代天文学研究的基础。

Along with the tremendous increase in equipment has come the introduction of such revolutionary things as transistors and other solid-state devices that have made passible, for the first time, small-size, highly reliable equipment with very low power drains.

随着设备数量的巨大增长,人们研制出了像晶体管的和其他固态器件这样革命性的东西,他们首次使制造出体积小、可靠性高、功耗很低的轻便设备成为可能。

(3)"分词+助动词 be 的时态形式+主语"的句型:

Shown on this page is a block diagram of a countercurrent solvent extraction.

这一页上画出的是逆流萃取的方框图。

Attached to one end of a rod is a body whose mass is twice that of the rod.

系在棒一端的是其质量为该棒质量两倍的一个物体。

Also omitted, to prevent the book from becoming unduly long, are discussions of design procedures and experimental techniques.

为了防止本书内容过于庞杂,还省去了对设计步骤和实验方法的讨论。

Distributed about the nucleus and revolving about it in orbits are much less massive negatively charged particles called electrons.

分布在原子核周围并在轨道上绕它旋转的是质量要比它小得多的称为电子的带负电的微粒。

Connecting the the cell to the rest of the control systen is the fiber cord.

使电解槽与其他控制系统相连的是光纤电缆。

(4)"介词短语(或引导词 there)+被动态谓语+主语"的句型:

In the central part of the glass electrolytic cell is located a set of electrodes known as the "electron gun".

在这玻璃电解槽的中心放置了一组称为电子枪的电极。

By a family of curves is meant a specified set of curves which satisfy given conditions.

所谓曲线族,指的是能满足给定条件的一组特殊的曲线。

A stereoscope is an instrument through which can be seen two pictures of the same scene.

立体视镜是通过它能看到同一景物的两种图像的一种仪器。

There is set up a transverse strain in addition to the longitudinal strain.

除了纵向应力外也产生了横向应力。

2.9.4.2 部分倒装

(1)"only+状语"开头的句型:

Only in such a case is dust comes out of the smelter.

只有在这种情况下,烟尘才从熔炼炉中逸出。

Only adding to catalyst, can the alternating leaching rate with little energy loss.

只是因为添加了催化剂,浸出率才能通过较小的能量调节。

Only then shall we take this measure.

只有在那时我们才采取这一措施。

Only under a matched condition is there a maximum leaching rate.

只有在合适条件下才会获得最大的浸出率。

(2)否定性副词或具有否定意义的介词短语处于句首作谓语的状语时的句型。常见的否定性副词或具有否定含义的介词短语有以下一些:

| | | | |
|---|---|---|---|
| never | 从不 | by no means | 绝不 |
| hardly | 几乎不 | in no way | 绝不 |
| scarcely | 几乎不 | at no time | 绝不 |
| seldom | 很少,难得 | in no case | 绝不 |
| rarely | 很少,难得 | on no account | 绝不 |
| little | 一点也不,毫不 | under no conditions | 绝不 |
| not always | 不总是 | under no circumstances | 绝不 |
| not only | 不仅 | on no condition | 绝不 |
| not until | 直到…… | | |

Not always does the addition or removal of heat to or from a sample of matter lead to a change in its temperature.

给一个物体增加热量或从该物体带走热量并不总会导致其温度的变化。

Not only are we able to find values of the functions if we know the angle, but also we can find the angle if we know the value of a function.

如果我们知道角度,我们不仅能求出各函数值来;而且我们若知道某个函数值,我们还能求出角来。

By no means can electrons move from the plate to the cathode.

电子绝不能从极板跑向阴极。

(3)有 so 或 neither/nor 处于句首的句型:

All elements do not necessarily function simultaneously, nor do they require similar information from the environment.

并非所有的要素一定同时起作用,它们也并不需要来自外界的相似信息。

Two electrons will be repelled from each and so will two nuclei.

两个电子相互排斥，两个原子核也是如此。

The electrical system in an automobile and an airplane uses the direct current and so do the telegraph and the telephone.

汽车和飞机中的电气系统采用的是直流电，而电报、电话也是如此。

In the absence of friction, the driving wheel would not run the belt, neither would the belt run the wheel to be driven.

若没有摩擦，主动轮就带不动皮带，而皮带也带不动从动轮。

(4)“宾语+主语+谓语”句型：

This solvent extraction process we shall discuss in detail.

这一萃取工艺我们将要详细讨论。

The other line we place vertically, and label the $y$-axis.

另一根线我们则将它放垂直，并把它标记为 $y$ 轴。

White dwarfs are probably very plentiful, but most of them we will never see.

白矮星很可能非常多，不过它们中大多数我们是永远也看不到的。

This we do know: these basic positive and negative charges are two of the basic building blocks from which the atom is constructed.

下面这一点我们确实是知道的：这些基本的正负电荷是构成原子的基本构件中间的两种。

(5)“状语或表语+主语+动词”的句型：

The individuals concerned are too numerous to mention but to all of them the author expresses his appreciation.

有关的个人太多了，无法一一提及，编者在此一并致谢。

With pressure the gaps decrease.

间隙随着压力的加大而减小。

Certain it is that all essential processes of hydrometallurgy occur in solvent.

确定无疑的是，湿法冶金的过程发生在溶剂中。

(6)“主语+谓语+宾补+宾语”句型：

Such oxidizing agent has the advantage of improving leaching rate.

这类氧化剂的优点是能够提高浸出率。

Friction makes necessary a good lubrication system.

摩擦使得有必要具备一个良好的润滑系统。

We can take as a second example the case shown in Fig. 5.

我们可以把图 5 所示的情况作为第二个例子。

We call insulators these substances which prevent the passage of electricity.

我们把阻止电通过的那些物质称为绝缘体。

This equivalent resistance could leave unaltered the potential difference between the terminals of the combination and the current in the rest of the circuit.

这个等效电阻能使该电阻组合两段的电位差以及电路其余部分的电流保持不变。

### 2.9.5 句子成分的省略

有时为了英语表达得更加精简，会常用到省略。常见的省略情况有以下几种：

（1）并列句的省略。并列句中，后一个分句与前一个分句相同的部分可以省略。如：

Matter consists of molecules and molecules of atoms.（在第二个 molecules 之后省去了 consist 一词）

物质是由分子构成的，而分子则是由原子构成的。

The earth attracts the moon and the moon the earth.（在第二个 the moon 之后省去了 attracts 一词）

地球吸引月球，月球吸引地球。

The meter is the standard for length, the gram for weights.（在 the gram 之后省去了 is the standard）

米是长度的度量标准，而克是质量的度量标准。

It is customary to consider $x$-components which are directed toward the right as positive and those toward the left, negative.（在 those 后省去了 which are directed，而在 left 后省去了 as）

通常我们把指向右边的 $x$ 分量看作为正，而把指向左边的 $x$ 分量看作为负。

Electrons are injected into the P region, and holes into the N region.（在 holes 后省去了 are injected）

电阻被注入 P 区，而空穴被注入 N 区。

Distances measured from an axis to the right or upward are positive; to the left or downward are negative.（在 to the left 之前省去了 distances measured from an axis）

在轴的右边和上方所测得的距离为正；在轴的左边和下方所测得的距离为负。

（2）状语从句中的省略。状语从句连接词+形容词/分词/介词短语/名词（一般省去 it is 或 they are）：

These gold particles, when heated, can emit electrons.（在 when 之后省去了 it is）

这些金粒子受热时能发射电子。

Every test, no matter how carefully made, is subject to experimental error.（在 made 之前省去了 it is）

每一个试验，不论做得如何仔细，均有实验性误差。

Although derived for the special cases of constant velocity and constant acceleration, the equations above are true in general.（在 derived 之前省去了 they were）

上述等式虽然是匀速及匀加速的特殊情况推导出来的，但它们在一般情况下均是适用的。

When pure, water is a colorless liquid.

水纯净时是一种无色的液体。

The acting and reacting forces in Fig. 1, though equal in magnitude and opposite in direction, can never neutralize each other because they always act on different objects.

图 1 中的作用力和反作用力虽然大小相等方向相反，但它们永远不可能相互抵消，因为它们总是作用在不同的物体上。

A constant number, however large, is never spoken of as infinite. （在 large 之后省去了 it is）

一个常数无论多大，绝不能说成是无穷大。

Most, if not all, communication circuits are available as integrated circuits. （句中的 not all 是作其后面名词的定语的，它与 most 一词的作用是一样的）

大多数（即使不是所有的）通信电路已有集成块了。

（3）两个介词共用介词宾语的情况。由于有时要求使用不同介词，为了精炼句子，可省去第一个介词的宾语。如：

It is necessary to find out the current through and the voltage across this resistor insert in the electrowinning cell.

必须求出这个插入电积槽内的电阻上的电流和电压。

Further inhalation of or even exposure to smog should be avoided.

必须避免更多地吸入烟雾，甚至应避免暴露在烟雾里。

In the measurement of microwave power the problem is to determine the actual amount of RF energy delivered to and used by the load.

在测量微波功率时，问题在于要确定馈给负载并由负载使用的射频能量的实际总量。

The accuracy of the location depends on the distance from and the direction of the shore stations.

定位的精度取决于离海岸电台的距离，同时取决于海岸电台的方向。

Analysis of convective heat transfer requires detailed knowledge of fluid motion in the presence of and adjacent to solid surfaces.

分析对热流传递需要详尽地了解存在固态表面时的流体运动情况以及邻近于固体表面的流体运动情况。

The following propositions, although incorrect( in part) in normal algebra, are correct in and basic to Boolean algebra.

下面这些命题虽然在普通代数中是部分不正确的，但是在布尔代数中是正确的，且是布尔代数的基础。

We wish to find out the product of the smelter.

我们想要知道熔炼炉中的产物。

The acceleration of a body is proportional to and in the direction of external forces acting upon it.

物体的加速度与作用在它上面的外力成正比，并与外力方向一致。

（4）由 hence 开头的省略句。如：

Stirring is essential for ore particle to contact with oxidizing agent adequately—hence their importance.

搅拌对矿物颗粒充分接触氧化剂十分必要，因而极为重要。

Semiconductors are a class of elements whose electrical properties lie in an area between conductors and insulators, hence their name.

半导体是其电性能处于导体和绝缘体之间某个范围内的一类元素，因而得到了其名称。

The stream of electrons is emitted from the cathode of the electron gun, hence the name "cathode ray".

电子流是从电子枪的阴极发射出来的，因而得到了“阴极射线”这一名称。

When the blood becomes viscous, it is difficult for the heart to pump it through the capillaries. Hence the increase in blood pressure.

当血液变黏后，心脏就很难把血液泵压通过毛细血管，因而血压就升高了。

### 2.9.6 句子成分的分隔

所谓的分隔，主要目的是表达上更加地道，以下举例说明：

（1）“主语+谓语+主语修饰语”句型。这一结构的主要目的为了避免句子发生头重脚轻的现象，其句型为：主语+谓语（不及物动词/及物动词被动语态/系动词+表语）+主语修饰语（定语从句/介词短语/同位语从句/现在分词短语/不定式短语）。如：

We assume that conditions exist for high leaching rate to occur.

我们假设存在着能产生高浸出率的条件。

In chapter 8, a cell design procedure is developed which is the reverse process of the analytical technique.

第 8 章讲解电解槽设计步骤，它是分析方法的逆过程。

The problem often arises of dividing one polynomial by another.

往往出现这样的问题：要用一个多项式除以另一个多项式。

Toward the end of the 19th century, evidence began to appear that atoms were not the ultimate particles of the universe.

到 19 世纪末才有证据表明：原子并不是宇宙间不能再分隔的粒子。

No charges have ever been found of smaller magnitudes than those of a proton or an electron.

至今尚未发现哪个电荷量比质子或电子所带的电量更小。

The theory is of great importance that the hotter a body is, the more energy it radiates.

物体越热，其辐射的能量越多，这一理论极为重要。

（2）名词后跟有两个修饰语的情况。第二个修饰语为 of 短语时，经常见到的分隔现象如下：

There are a number of other striking examples that can be given which bear upon heat and temperature.

我们可以举出另外一些与热和温度有关的、引人注目的例子来。

Electronic measurements are of two kinds: those made of electric quantities such as voltage, capacitance, or field strength, and those made by electronic means of other quantities such as pressure, temperature, or flow rate.

电子测量分两种：一种是对像电压、电容或场强这些电子量所进行的测量，另一种是对其他一些量如压力、温度或流速用电子方法所进行的测量。

科技英语写作中，有一些固定的表达可以直接使用，对于初学者而言，适当地使用一些句型不失为一种好办法。

表达可能性：

It is certain that...

It is almost certain that...

It is very probable/likely that...

It is possible that...

It is unlikely that...

It is very/highly unlikely...

There is a definite possibility that...

There is a strong possibility that...

There is a good possibility that...

There is a slight possibility that...

There is little possibility that...

表示例外：

With the exception of...

Apart from...

Except for...

表示原因：

be due to...

be attributed to...

be accounted for...

be a consequence of...

stem from

表示定义：

be named...

be called...

be denoted by...

be known as...

be defined as...

be referred to...

引出设问句：

It is uncertain/unclear...

It has not been determined...

It is necessary to consider...

The question remains...

We need to know/consider...

指出问题：

This system/process/idea has its problems.

There remains the issue of reliability.

This model has some serious limitations.

Few solutions have been found to...

Little progress has been made in...

The problem remains as to how...

Researchers still have to find a way to...

引出话题：

Recently, there has been growing interest in...

The possibility of... has generated wide interest in...

The development of... is a classic problem in...

The development of... has led to the hope that. .

The... has become a favorite topic for analysis...

Knowledge of... has great importance for...

The study of... has become an important aspect of...

A central issue in... is...

(The) ... has been extensively studied in recent years.

Many investigators have recently turned to...

The relationship between... and... has been investigated by many researchers.

Many recent studies have focused on...

指出研究的局限：

It should be noted that this study has been primarily concerned with...

This analysis has concentrated on…

The findings of this study are restricted to…

This study has addressed only the question of…

The limitation of this study is clear: …

We would like to point out that we have not…

表达研究的意义：

Notwithstanding its limitations, this study does suggest…

Despite its preliminary character, the research reported here would seem to indicate…

However exploratory, this study may offer some insight into…

## 2.10 有色冶金常用英语词汇表

有色冶金常用英语词汇见表2-2。

**表2-2 有色冶金常用英语词汇**

| 英 文 | 中 文 | 英 文 | 中 文 |
|---|---|---|---|
| abrasion resistance | 耐磨性 | bronze | 青铜 |
| absorption | 吸收 | bubbling | 鼓泡 |
| acid oxide | 酸性氧化物 | calcine | 焙砂 |
| activation energy | 活化能 | carbon anode | 碳阳极 |
| activity | 活度 | carbon cathode | 碳阴极 |
| activity coefficient | 活度系数 | cast rolling directly from liquid metal | 无锭轧制 |
| adiabatic process | 绝热过程 | cathodic copper | 阴极铜 |
| agglomeration | 团聚 | chain reaction | 链反应 |
| alumina | 氧化铝 | chemical equilibrium | 化学平衡 |
| aluminum alloy | 铝合金 | chemical kinetics | 化学动力学 |
| aluminum hydroxide | 氢氧化铝 | chemical potential | 化学位 |
| aluminum hydroxide/hydrate | 氢氧化铝 | chemical process | 化学过程 |
| anode plate | 阳极板 | chemical reaction | 化学反应 |
| apparent activation energy | 表观活化能 | chemical reaction isotherm | 化学反应等温式 |
| Arrhenius equation | 阿伦尼乌斯方程 | chemical vapor deposition（CVD） | 化学气相沉积 |
| basicity of slag | 渣碱度 | coarse fraction | 粗粒级 |
| bauxite | 铝土矿 | coarse particle | 粗颗粒 |
| Bayer process | 拜耳法 | collision theory | 碰撞理论 |
| biocompatibility | 生物相容性 | combination reaction | 化合反应 |
| blister/crude copper | 粗铜 | concentrate | 精矿 |
| boundary layer | 边界层 | concentration | 富集 |
| brass | 黄铜 | concentration of solution | 溶液浓度 |
| brittleness | 脆性 | continuous casting-direct rolling | 连铸连轧 |

续表 2-2

| 英　文 | 中　文 | 英　文 | 中　文 |
|---|---|---|---|
| copper alloy | 铜合金 | ferromagnetism | 铁磁性 |
| copper concentrate | 铜精矿 | Fick's 1st law of diffusion | 菲克第一扩散定律 |
| copper sulfide | 硫化铜 | Fick's 2nd law of diffusion | 菲克第二扩散定律 |
| corrosion resistance | 耐腐蚀性 | fine fraction | 细粒级 |
| crude | 原矿 | fine particle | 细颗粒 |
| crushing | 破碎 | first order reaction | 一级反应 |
| cryolite | 冰晶石 | fusion cast | 熔铸 |
| cupronickel | 白铜 | gas bubble | 气泡 |
| cutting | 切削加工 | Gibbs energy | 吉布斯能 |
| decomposition reaction | 分解反应 | Gibbs energy of formation | 生成吉布斯能 |
| demanganization | 脱锰 | Gibbs energy of reaction | 反应吉布斯能 |
| density | 密度 | Gibbs-Helmholtz equation | 吉布斯-亥姆霍兹方程 |
| deoxidation constant | 脱氧常数 | graphite | 石墨 |
| desiliconization | 脱硅 | grinding | 磨碎 |
| desulfurizer | 脱硫剂 | half-life | 半衰期 |
| digestion | 溶出 | hardness | 硬度 |
| dispersed metal | 分散金属 | heap/dump leaching | 堆浸 |
| displacement reaction | 置换反应 | heat capacity | 热容 |
| distribution law | 分配平衡 | heat effect | 热效应 |
| electric conductivity | 导电性 | heat of fusion | 熔化热 |
| electro-deposited copper | 电积铜 | heat of phase transformation | 相变热 |
| electrolytic copper | 电解铜 | heat of sublimation | 升华热 |
| electrolyzer | 电解槽 | heat of vaporization | 汽化热 |
| electrometallurgy | 电冶金学 | heat transfer | 传热 |
| elementary reaction | 基元反应 | heavy non-ferrous metal | 重有色金属 |
| endothermic reaction | 吸热反应 | homogeneous system | 均相系统 |
| enthalpy | 焓 | hydrometallurgy | 湿法冶金学 |
| enthalpy of formation | 生成焓 | ideal solution | 理想溶液 |
| enthalpy of reaction | 反应焓 | in-situ leaching | 原地浸出 |
| entropy | 熵 | integral molar quantity | 总物质的量 |
| equilibrium | 平衡 | intensive property | 强度性质 |
| equilibrium constant | 平衡常数 | interaction coefficient | 相互作用系数 |
| equilibrium value | 平衡值 | irreversible process | 不可逆过程 |
| exothermic reaction | 放热反应 | irreversible reaction | 不可逆反应 |
| extensive property | 广度性质 | isobaric process | 等压过程 |
| extracting | 提取 | isochoric process | 等容过程 |

续表 2-2

| 英 文 | 中 文 | 英 文 | 中 文 |
|---|---|---|---|
| isothermal process | 等温过程 | nonferrous metallurgy | 有色冶金学 |
| jet | 射流 | *n*th order reaction | *n* 级反应 |
| Kelvis（K） | 开氏温度 | ore | 矿石 |
| kinetics of metallurgical process | 冶金过程动力学 | ore grade | 矿石品位 |
| laminar flow | 层流 | overall reaction | 总反应 |
| law of mass action | 质量作用定律 | oversaturated | 过饱和 |
| light metal | 轻金属 | parallel reaction | 平行反应 |
| liquid droplet | 液滴 | partial molar quantity | 偏摩尔量 |
| load-bearing structure | 承力件 | particle size | 粒度 |
| macrokinetics | 宏观动力学 | phase diagram | 相图 |
| magnesium alloy | 镁合金 | phase equilibrium | 相平衡 |
| magnetostriction | 磁致伸缩性 | physical process | 自理过程 |
| mass transfer | 传质 | polishing | 抛光 |
| mechanical property | 力学性能 | precision alloy | 精密合金 |
| melting point | 熔点 | primary aluminum | 原铝 |
| mesh | 网目 | pure substance standard | 纯物质标准态 |
| metalloid | 类金属 | pyrometallurgy | 火法冶金学 |
| metallothermic reduction | 金属热还原 | quartz | 石英 |
| metallurgical engineering | 冶金工程 | radioactive metal | 放射性金属 |
| metallurgical process | 冶金过程 | radioactivity | 放射性 |
| metallurgy | 冶金学 | rare earth metal | 稀土金属 |
| microkinetics | 微观动力学 | rare metal/scarce metal | 稀有金属 |
| mineral | 矿物 | reaction mechanism | 反应机理 |
| mineral deposit | 矿床 | reaction order | 反应级数 |
| mining | 采矿 | reaction rate | 反应速率 |
| mole fraction | 摩尔分数 | reaction rate constant | 反应速率常数 |
| molten salt | 熔盐 | real solution | 真实溶液 |
| momentum transfer | 动量传输 | recovery | 回收率 |
| multicomponent system | 多元系 | refining | 精炼 |
| multiphase reaction | 多相反应 | reflectivity | 反射性 |
| Ni-base superalloy | 镍基高温合金 | regular solution | 正规溶液 |
| nickel alloy | 镍合金 | resistance | 电阻 |
| noble metal | 贵金属 | resistance to nuclear radiation | 耐核辐射 |
| nodular cast iron | 球墨铸铁 | reversible process | 可逆过程 |
| nodulizer | 球化剂 | reversible reaction | 可逆反应 |
| non-ferrous alloy | 有色合金 | rigidity | 刚性 |

续表 2-2

| 英　文 | 中　文 | 英　文 | 中　文 |
|---|---|---|---|
| second order reaction | 二级反应 | tailings | 尾矿 |
| selective oxidation | 选择性氧化 | temperature coefficient | 温度系数 |
| semiconductor | 半导体 | the Periodic Table of Chemical Elements | 化学元素周期表 |
| semimetal | 半金属 | thermal conductivity | 导热性 |
| separation | 分离 | thermochemistry | 热化学 |
| shape memory | 形状记忆 | thermodynamic equilibrium | 热力学平衡 |
| sieving | 筛分 | thermodynamic function | 热力学函数 |
| single-component system | 单元系 | thermodynamics of metallurgical processes | 冶金过程热力学 |
| sodium aluminate solution | 氯酸钠溶液 | titanium alloy | 钛合金 |
| soft magnetic alloy | 软磁合金 | toughness | 韧性 |
| solid phase | 固相 | transport phenomena | 传输现象 |
| solid solution | 固溶体 | turbine blade | 涡轮叶片 |
| solute | 溶质 | turbulent flow | 湍流 |
| solvent | 溶剂 | ultrafine particle | 超微颗粒 |
| sound absorption property | 吸声性 | vacuum metallurgy | 真空冶金学 |
| spontaneous process | 自发过程 | valuable mineral | 有用矿物 |
| stainless steel | 不锈钢 | viscosity | 黏度 |
| standard state | 标准态 | wrought aluminum alloy | 变形铝合金 |
| statistical thermodynamics | 统计热力学 | yield | 产率 |
| strength | 强度 | zero order reaction | 零级反应 |
| strength to weight ratio | 比强度 | | |

# 3 有色冶金英文文献检索

## 3.1 文献检索方法

通常而言，检索文献和阅读文献的时间应该占到一个研究项目总时间的一半左右。从选题、确定实验方案、实验仪器及材料的选择，到理论分析、撰写论文，各个方面都需要检索和阅读文献。阅读优质文献的好处有减少重复的工作量，在别人实验结果的基础上再深入研究。

不同的科研类型对信息检索的要求也不同。对于基础研究，推荐用 SCI 检索。在选择文献时，应重点选择影响因子高，引用次数多的文章。对于应用研究而言，SCI、EI、CA 和专利数据库都可使用。此外，中文科技期刊如知网、维普和万方等也是不错的数据库。以下介绍两种常用的文献检索手段。

### 3.1.1 SCI 检索

通常我们使用 Web of Science 中 SCI 数据库进行 SCI 检索。Web of Science 整合了 SCI（Science Citation Index），SSCI（Social Science Citation Index），AHCI（Arts & Humanities Citation Index）等数据库，利用互联网创建了多学科、多领域的文献数据库。选择参考文献时常参考的指标影响因子（Impact Factor，IF）为该网站推出，目前已经成为国际上通用的评价指标。其网站页面如图 3-1 所示。

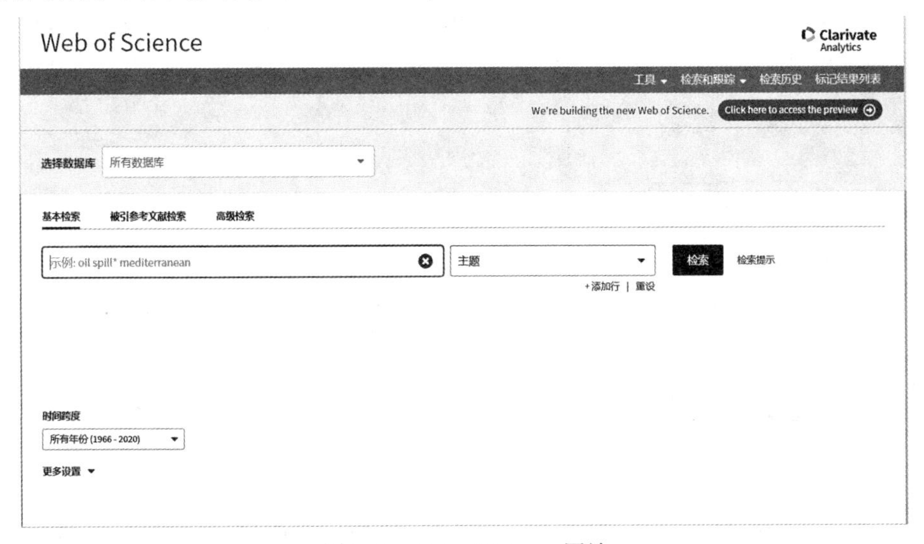


图 3-1 Web of Science 网站

检索文献前需要选择数据库，以达到更好的检索效果。Web of Science 的数据库如图 3-2所示。

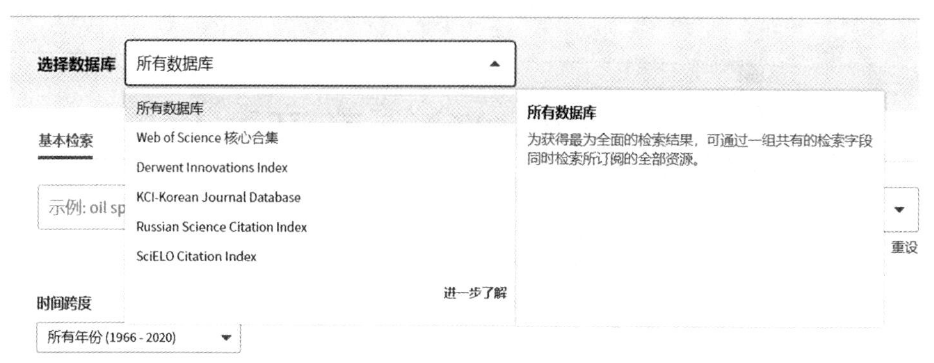


图 3-2　Web of Science 数据库

选择数据库后，还可以有针对性地检索部分内容。如文献的主题、标题、作者、出版物名称或出版年等，如图 3-3 所示。

图 3-3　检索图

如想要检索 extraction of rare earths 相关内容，按照主题检索后，根据被引频次排序，选择被引频次最高的文献。检索结果右侧可以看到每篇文章的被引频次，如图 3-4 所示。

排序方式: 日期　被引频次　使用次数　相关性　更多　　1 / 129

选择页面　导出...　添加到标记结果列表　　分析检索结果　创建引文报告

1. A critical review on solvent extraction of rare earths from aqueous solutions
作者: Xie, Feng; Zhang, Ting An; Dreisinger, David; 等.
MINERALS ENGINEERING 卷: 56 页: 10-28 出版年: FEB 2014
Context Sensitive Link　出版商处的免费全文　查看摘要
被引频次: 499 (来自所有数据库)　高被引论文　使用次数

2. Rare Earth Elements: Minerals, Mines, Magnets (and More)
作者: Chakhmouradian, Anton R.; Wall, Frances
ELEMENTS 卷: 8 期: 5 页: 333-340 出版年: OCT 2012
Context Sensitive Linl　出版商处的全文　查看摘要
被引频次: 257 (来自所有数据库)　高被引论文　使用次数

3. Towards zero-waste valorisation of rare-earth-containing industrial process residues: a critical review
作者: Binnemans, Koen; Jones, Peter Tom; Blanpain, Bart; 等.
JOURNAL OF CLEANER PRODUCTION 卷: 99 页: 17-38 出版年: JUL 15 2015
Context Sensitive Linl　出版商处的全文　查看摘要
被引频次: 232 (来自所有数据库)　高被引论文　使用次数

(a)

Web of Science

Clarivate Analytics

检索 返回检索结果 工具 检索和跟踪 检索历史 标记结果列表

出版商处的免费全文 全文选项 导出... 添加到标记结果列表

第1条，共1,283条

A critical review on solvent extraction of rare earths from aqueous solutions

作者: Xie, F (Xie, Feng)[1]; Zhang, TA (Zhang, Ting An)[1]; Dreisinger, D (Dreisinger, David)[2]; Doyle, F (Doyle, Fiona)[3]

查看 Web of Science ResearcherID 和 ORCID

MINERALS ENGINEERING

卷: 56 页: 10-28

DOI: 10.1016/j.mineng.2013.10.021

出版年: FEB 2014

文献类型: Review

查看期刊影响力

引文网络

在 Web of Science 核心合集 中

491

被引频次

高被引论文

创建引文跟踪

(b)

图 3-4 检索结果

点击出版商处的免费全文，得到文献的下载界面，如图 3-5 所示。

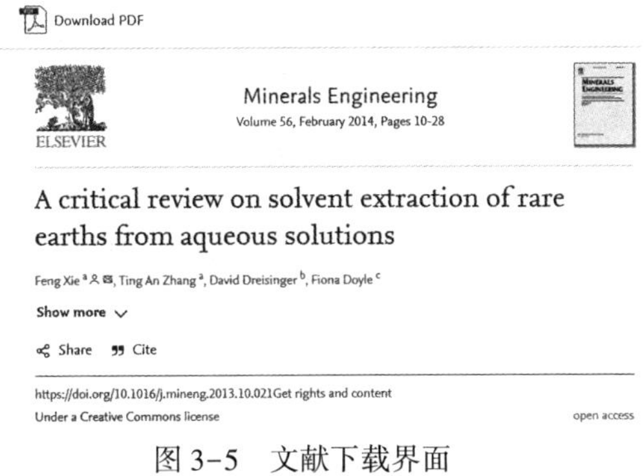

Download PDF

ELSEVIER

Minerals Engineering

Volume 56, February 2014, Pages 10-28

A critical review on solvent extraction of rare earths from aqueous solutions

Feng Xie [a], Ting An Zhang [a], David Dreisinger [b], Fiona Doyle [c]

Show more

Share Cite

https://doi.org/10.1016/j.mineng.2013.10.021Get rights and content

Under a Creative Commons license open access

图 3-5 文献下载界面

点击“Download PDF”，得到文献全文，如图 3-6 所示。

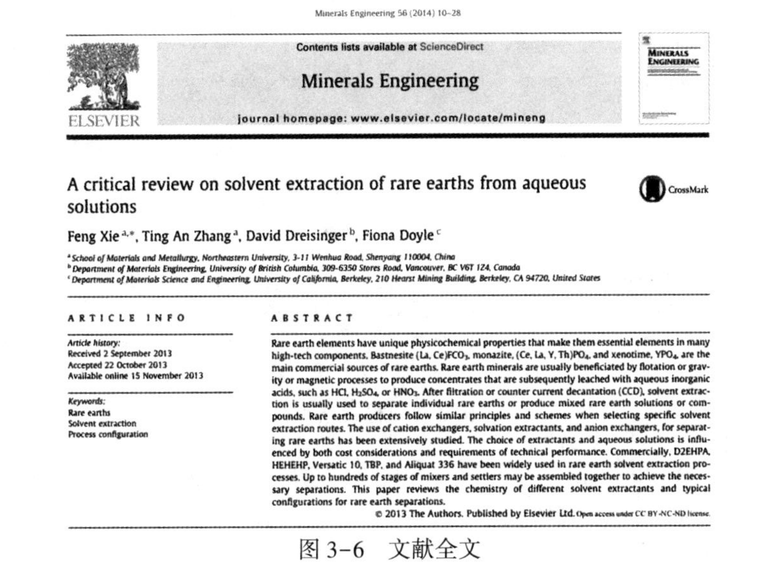

Minerals Engineering 56 (2014) 10-28

Contents lists available at ScienceDirect

Minerals Engineering

journal homepage: www.elsevier.com/locate/mineng

ELSEVIER

A critical review on solvent extraction of rare earths from aqueous solutions

CrossMark

Feng Xie [a,*], Ting An Zhang [a], David Dreisinger [b], Fiona Doyle [c]

[a] School of Materials and Metallurgy, Northeastern University, 3-11 Wenhua Road, Shenyang 110004, China

[b] Department of Materials Engineering, University of British Columbia, 309-6350 Stores Road, Vancouver, BC V6T 1Z4, Canada

[c] Department of Materials Science and Engineering, University of California, Berkeley, 210 Hearst Mining Building, Berkeley, CA 94720, United States

ARTICLE INFO

*Article history:*
Received 2 September 2013
Accepted 22 October 2013
Available online 15 November 2013

*Keywords:*
Rare earths
Solvent extraction
Process configuration

ABSTRACT

Rare earth elements have unique physicochemical properties that make them essential elements in many high-tech components. Bastnesite (La, Ce)$FCO_3$, monazite, (Ce, La, Y, Th)$PO_4$, and xenotime, $YPO_4$, are the main commercial sources of rare earths. Rare earth minerals are usually beneficiated by flotation or gravity or magnetic processes to produce concentrates that are subsequently leached with aqueous inorganic acids, such as HCl, $H_2SO_4$, or $HNO_3$. After filtration or counter current decantation (CCD), solvent extraction is usually used to separate individual rare earths or produce mixed rare earth solutions or compounds. Rare earth producers follow similar principles and schemes when selecting specific solvent extraction routes. The use of cation exchangers, solvation extractants, and anion exchangers, for separating rare earths has been extensively studied. The choice of extractants and aqueous solutions is influenced by both cost considerations and requirements of technical performance. Commercially, D2EHPA, HEHEHP, Versatic 10, TBP, and Aliquat 336 have been widely used in rare earth solvent extraction processes. Up to hundreds of stages of mixers and settlers may be assembled together to achieve the necessary separations. This paper reviews the chemistry of different solvent extractants and typical configurations for rare earth separations.



图 3-6 文献全文

### 3.1.2 知网检索

知网的概念是国家知识基础设施（National Knowledge Infrastructure，NKI），由世界银行于 1998 年提出。CNKI 工程是以实现全社会知识资源传播共享与增值利用为目标的信息化建设项目，由清华大学、清华同方发起，始建于 1999 年 6 月。知网的首页如图 3-7 所示。同样的我们可以选择不同的检索内容进行有针对性的检索。

图 3-7 知网首页

如检索稀土元素的提取相关内容，可以检索主题“稀土 提取”，得到检索结果如图 3-8 所示，按被引次数排序。

图 3-8 检索结果

选择被引频次第一的文章，打开出现如图 3-9 所示的界面。

点击底部的 PDF 下载，可得到文献全文，如图 3-10 所示。

中国稀土学报．2007年06期　第641-650页　北大核心

记笔记

## 风化壳淋积型稀土矿评述

池汝安　田君

武汉工程大学绿色化工过程省部共建教育部重点实验室湖北省新型反应器与绿色化学工艺重点实验室　中南大学资源加工与生物工程学院 湖北武汉430073　湖南长沙410083　江西科学院化学所　江西南昌330029

**摘要：** 对我国特有的风化淋积型稀土矿的科学研究和工业利用进行了评述,阐述了该矿成矿过程,探讨了稀土元素在风化体系中的迁移富集规律,证实稀土配分的铈亏效应、富铕效应、分馏效应和钆断效应,论述了风化壳淋积型稀土矿浸取水动力学、浸取动力学及浸取传质过程。提出了该矿稀土回收中存在的一些问题,建议进一步完善稀土元素地球化学和无机化学的科学体系,开发新浸取剂,特别是组合浸取剂的协同使用,深入研究原地浸出过程中的扩散和传质机理,强化浸取过程和建立相应的数学模型,有效地提高稀土浸出率,从而完善风化壳淋积型稀土矿的原地浸出理论和技术。

**关键词：** 风化壳淋积型稀土矿；浸出；矿物学；稀土；

**基金资助：** 国家自然科学基金资助项目(59674021,50474022,50664004,50574069)；国家杰出青年科学基金资助项目(59725408)；

**专辑：** 理工A(数学物理力学天地生)；理工B(化学化工冶金环境矿业)

**专题：** 地质学；矿业工程

**分类号：** P618.7

图3-9　文献下载界面

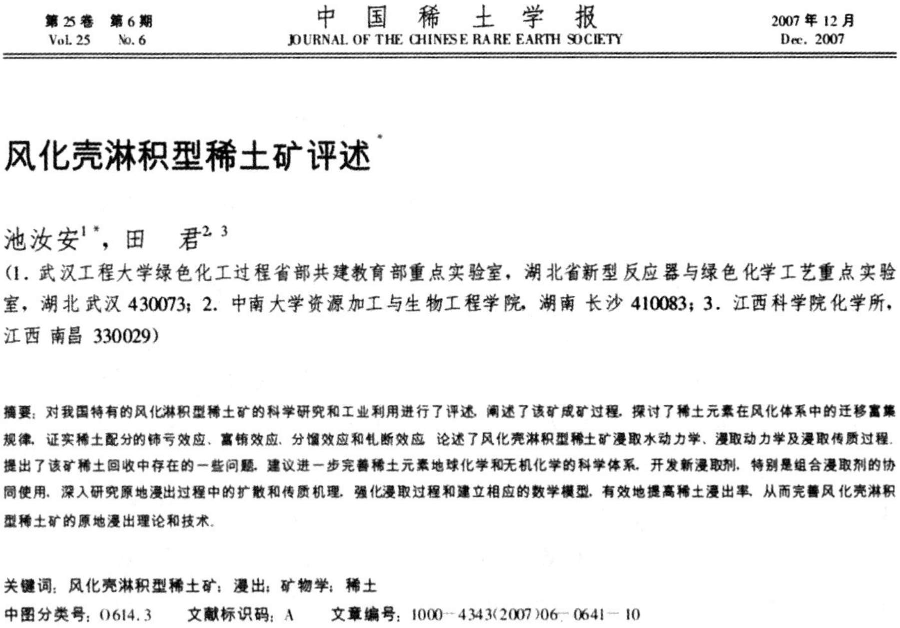

第25卷　第6期　　中　国　稀　土　学　报　　2007年12月
Vol. 25　No. 6　　JOURNAL OF THE CHINESE RARE EARTH SOCIETY　　Dec. 2007

# 风化壳淋积型稀土矿评述*

池汝安[1*]，田　君[2,3]

(1. 武汉工程大学绿色化工过程省部共建教育部重点实验室，湖北省新型反应器与绿色化学工艺重点实验室，湖北 武汉 430073；2. 中南大学资源加工与生物工程学院，湖南 长沙 410083；3. 江西科学院化学所，江西 南昌 330029)

摘要：对我国特有的风化淋积型稀土矿的科学研究和工业利用进行了评述．阐述了该矿成矿过程．探讨了稀土元素在风化体系中的迁移富集规律．证实稀土配分的铈亏效应、富铕效应、分馏效应和钆断效应．论述了风化壳淋积型稀土矿浸取水动力学、浸取动力学及浸取传质过程．提出了该矿稀土回收中存在的一些问题．建议进一步完善稀土元素地球化学和无机化学的科学体系．开发新浸取剂．特别是组合浸取剂的协同使用．深入研究原地浸出过程中的扩散和传质机理．强化浸取过程和建立相应的数学模型．有效地提高稀土浸出率．从而完善风化壳淋积型稀土矿的原地浸出理论和技术．

关键词：风化壳淋积型稀土矿；浸出；矿物学；稀土

中图分类号：O614.3　　文献标识码：A　　文章编号：1000－4343(2007)06－0641－10

图3-10　文献全文

检索时，我们通常会遇到检索结果过多或过少的问题。出现这种现象通常是由以下三个原因导致的：

（1）所研究领域的文献确实很多或很少；

（2）对课题背景了解不深，检索策略不当导致；

（3）不熟悉数据库的特点导致误检。

对于检索结果过多的现象，可以通过以下方法改进：

（1）避免使用过于宽泛的检索词，如直接用“湿法冶金”或“火法冶金”进行检索；

（2）检索避免使用具有多重含义的词语；

（3）检索的单词尽可能完整一些，如避免使用“Prep *”检索“Preparation”。

对于检索结果过少的现象，可以通过以下方法改进：

（1）避免使用不规范的关键词或俗称，如避免使用“水银”检索“汞”或使用“烧碱”检索“氢氧化钠”；

（2）避免使用不当的上位概念和下位概念。

## 3.2 文献管理及引用常用工具

随着科研工作的进行，浏览过的文献越来越多，这就使得文献的管理成了一个大问题。目前市面上常见的文献管理软件有几十种，其中最主流的有：Endnote，Mendeley 和 NoteExpress 等。这些文献管理软件可以帮助我们在写论文时管理引用参考文献，自动排序参考文献，还可以让我们快速检索到已经阅读过的文献，针对不同类型的文献进行分组等。下面主要介绍 Endnote 的使用。

### 3.2.1 Endnote

Endnote 是 Thomson Corporation 下属的 Thomson Research Soft 开发的一款文献管理软件（官网：http：//www.endnote.com/），是大多数科研工作者首先推荐的文献管理软件。其功能非常强大，可以解决绝大多数文献管理遇到的问题。Endnote 支持绝大多数国际期刊的参考文献格式。除了参考文献外，还提供不同期刊的写作模板。软件还提供 Word 插件，用 Microsoft Word 写作论文时，配合 Endnote 插件使用，基本上可以解决所有引用参考文献出现的问题。

本地的 Endnote 数据还可以同步到云备份上，只需要登录 Endnote 账户就可以在其他地方同步文献数据。软件本身还可以检索文献，其内嵌的数据库可以直接在软件中进行文献检索。

在 EndNote 中文帮助中（https：//www.howsci.com/endnote/），详细地列举了 Endnote 常见功能的用法。需要注意的是 EndNote 是一款收费软件，使用时有免费试用期可以体验。接下来简单介绍往 Endnote 中导入文献以及在 word 中引入参考文献的方法。

一般文献检索网站上都会有“Export”选项，以下以 Science Direct 为例，检索出目标文献后，可以看到“Export”选项，如图 3-11 所示。

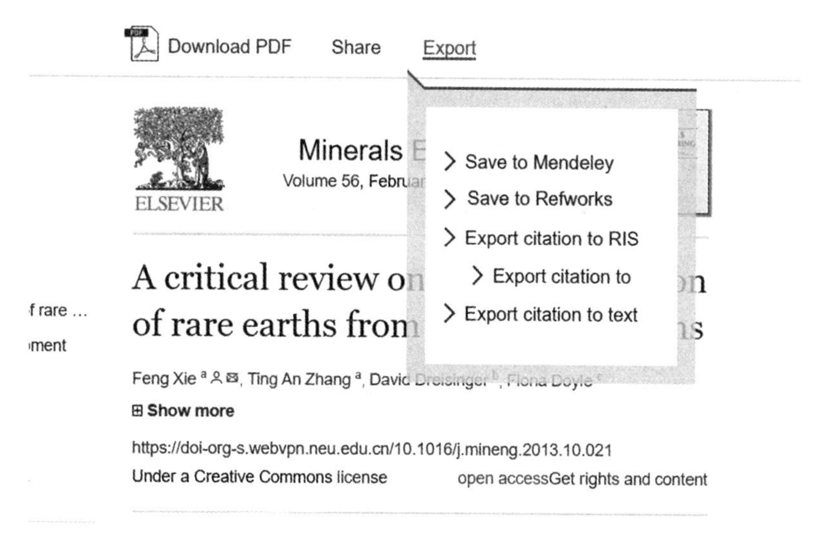


图 3-11　Export 选项图

点击“Export”后，选择“Export citation to RIS”，将会从网站上下载后缀为 . ris 格式的文件。用 Endnote 打开后，文献的信息会自动导入，如图 3-12 所示。

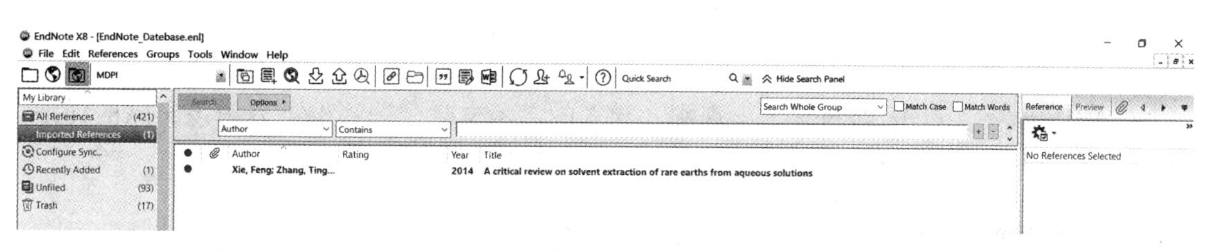


图 3-12　Endnote 导入文献

得到该文献信息后，可以在软件中对其进行分组管理。随后可以在文章中引用。引用有两种方法，第一种方法是直接在软件中选择文献，然后按下“Ctrl+C”即可复制，在文章中参考文献引用的目标位置处按“Ctrl+V”即可粘贴。第二种方法是在 Word 中选择 EndNote 插件，如图 3-13 所示。

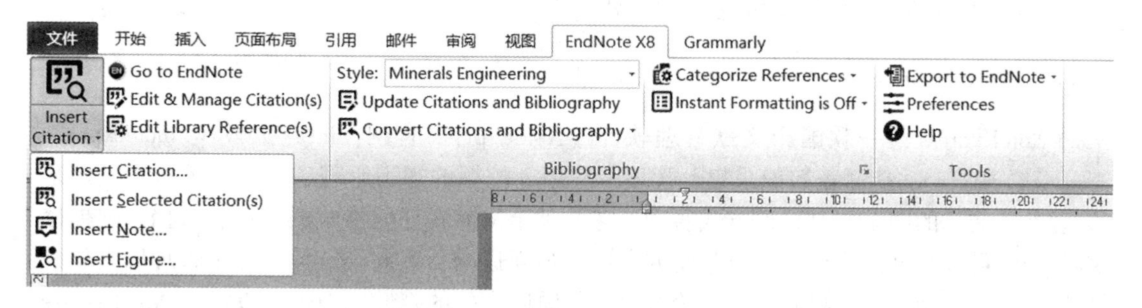


图 3-13　Word 中 Endnote 插入导入文献

选择最左侧的 Insert Citation 按钮，点击 Insert Citation，出现如图 3-14 所示的选项框。

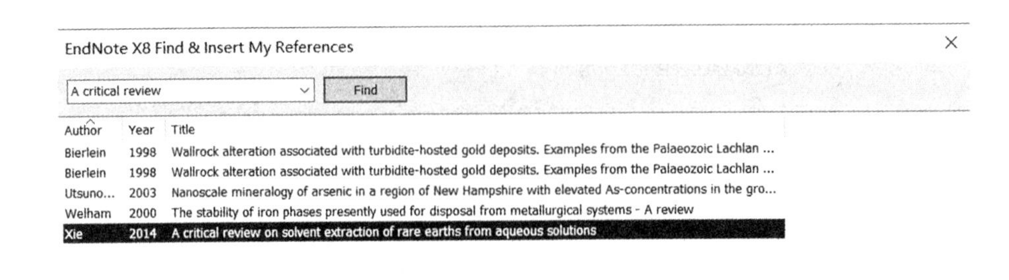


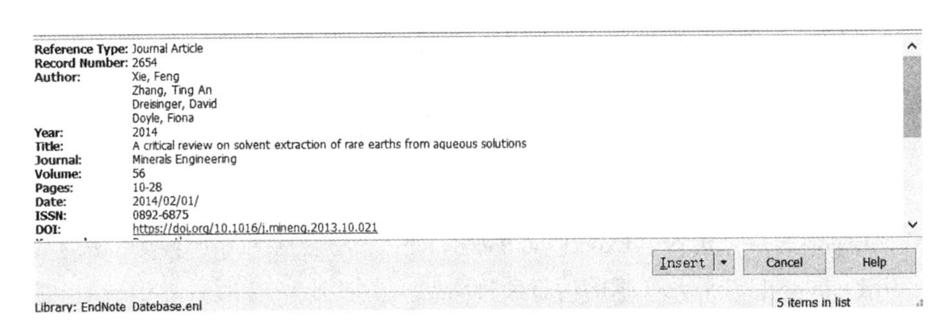


图 3-14 Insert Citation 选项框

输入关键字，点击“Find”，出现刚才导入的参考文献，选项框下方为该文献的详细信息。点击右下侧的“Insert”即可在论文中插入参考文献。

### 3.2.2 Mendeley

Mendeley 是一款免费的文献管理软件，同时也是一个在线的学术社交网络平台（Mendeley 官网：https://www.mendeley.com/dashboard/）。可一键抓取网页上的文献信息添加到个人的“library”中。可安装 Word 插件，方便在文字编辑器中插入和管理参考文献。Mendeley 免费提供各 2GB 的文献存储和 100MB 的共享空间。此外，Mendeley 加入了新的文献评价指标 Altmetric，能够反映新兴社交媒体的影响力。目前 Mendeley 已经被老牌科技出版巨头 Elsevier 收购，目前依旧可以免费使用。

### 3.2.3 NoteExpress

NoteExpress 是一款国产文献管理软件（NoteExpress 官网：http://www.inoteexpress.com/），其核心功能涵盖知识采集、管理、应用、挖掘的知识管理的所有环节。可以用来管理参考文献的题录，以附件方式管理参考文献全文或者任何格式的文件、文档。数据挖掘的功能可以帮助用户快速了解某研究方向的最新进展、各方观点等。除了管理以上显性的知识外，类似日记，科研心得，论文草稿等瞬间产生的隐性知识也可以通过 NoteExpress 的笔记功能记录，并且可以与参考文献的题录联系起来。需要注意的是这个软件是商业软件，目前永久版需要收费。

# 4 有色冶金英文论文典型格式

## 4.1 图片与表格

图片与表格一般用于呈现试验结果（如试验数据的陈列或绘制为曲线图、散点图等）。读者阅读文献时，在浏览过标题和摘要后，一般会接着浏览文章的图片和表格，使读者对作者的研究方向和研究内容有大致的了解。图片与表格能够比文字更直观地表述一些内容。大多数刊物都会规定图片和表格的格式，作者投稿时需注意阅读作者须知。但有一些要求是通用的，接下来分开介绍图片和表格的一些通用格式要求。

### 4.1.1 绘制图片的注意事项

绘制图片的注意事项包括：

（1）注明 $X$ 轴和 $Y$ 轴的单位；

（2）注意使用刻度单位；

（3）图片的标签、标尺、单位、标题等大小应便于阅读；

（4）图片表述的内容一定要清晰：

1）多条曲线用相同的标注方式；

2）多条曲线相互交错，曲线交汇处看不清；

3）在一张图中绘制多条曲线。

### 4.1.2 绘制表格的注意事项

大多数表格由以下几个部分构成。

首先是表的编号，不同刊物的编号方式略有差别，作者投稿时需注意。其次是表的标题。标题应该遵循简洁的原则，不需要包含表中结果的解释，详细描述等内容。如：

Table 1 Oxygen partial pressure of autoclave at 393K, 413K, 433K and 453K

应该改为：

Table 1 Oxygen partial pressure of autoclave

除了表格的编号和标题外，还有行标题和列标题。值得注意的是，行标题和列标题的数据如果有单位一定注明。表格中如果有缺省数据，可以用破折号（——）或者加空格的句点（…）标明。

有时候我们还需要对表格中的数据进行标注。标注表格中的数据需要用到脚注。脚注在表格的底部。不同期刊对脚注可能有要求。通常使用 *（星号）、†（单剑号）、‡（双剑号）、§（段落）和#（井字记号）。一个脚注要重起一行，脚注之间需加入一个空白行。此外，脚注的字号通常小于表格的字号。

下面以“Metals”期刊为例，该期刊对于表格和图片的要求如下：

All figures and tables should be cited in the main text as Figure 4.1, Table 4.1, etc.

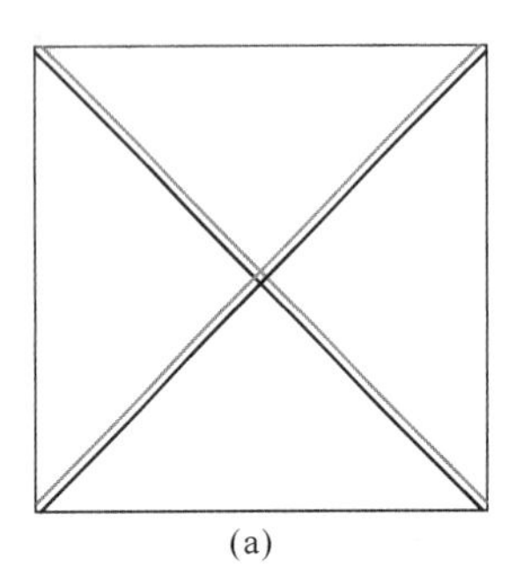
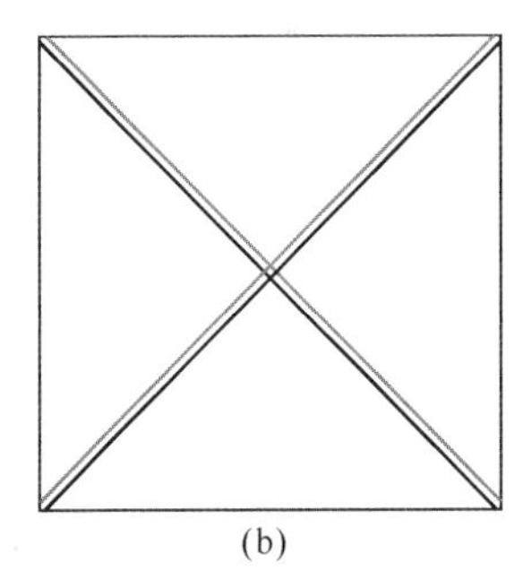

(a) (b)

Figure 4.1 This is a figure, Schemes follow the same formatting. If there are multiple panels, they should be listed as: (a) Description of what is contained in the first panel; (b) Description of what is contained in the second panel. Figures should be placed in the main text near to the first time they are cited. A caption on a single line should be centered.

**Table 4.1 This is a table. Tables should be placed in the main text near to the first time they are cited**

| Title 1 | Title 2 | Title 3 |
|---|---|---|
| entry 1 | data | data |
| entry 2 | data | data [1] |

[1] Tables may have a footer.

## 4.2 标点符号

标点符号是书面上用于标明句读和语气的符号，能够辅助文字记录语言，是书面语的组成部分，用来表示停顿、语气以及词语的性质和作用。在英文中，对于标点符号的使用具有一些默认的规则，学习英文写作时，务必要了解英文标点符号的使用。本节主要介绍英文标点符号的使用规则以及一些常见的错误。

（1）逗号（comma）。逗号的常规用法是句子之间的停顿符号。以下列举几种常见逗号的用法：

1）区分并列连词所连接的分句：

I'm sorry, but it's nothing to do with you.

You could only really tell the effects of the disease in the long term, and five years wasn't long enough.

2）复合句主句和简单句前的从句后，通常会用逗号标示：

When people, animals, and plants grow, they increase in size and change physically over a period of time.

3）逗号可以用于隔出非限定修饰词语。非限定性修饰词语提供补充性的信息，可以省略而不改变句子的原有含义。而限定修饰词语如果被省略，则会改变句子原有意思，因此是不能省略的。如：

These apple trees, which I planted three years ago, have not borne any fruit. She was very patient towards the children, which her husband seldom was.

When deeply absorbed in work, which he often was, he would forget all about eating and sleeping.

The textile mill has over 8, 000 workers and staffs, eighty per cent of whom are women.

4）主句和独立短语之间需要用逗号隔开。从语法上看，独立短语可视为状语，修饰整个句子：

The work having been done, the stuff began their rests.

5）使用插入语时，应该用逗号隔出。插入语是指在一个句子中间插入的一个成分，它不做句子的何种成分，也不和句子的何种成分发生结构关系，同时既不起连接作用，也不表示语气的语段：

The first thing a right-minded girl must do, if only for a second, is take a look around her.

We cannot, as much as I think it is right and just that we should, have our cakes and eat them too.

I really was, unbeknown to myself, deeply despondent.

6）使用逗号分开三个或三个以上的并列词、短语或从句：

High temperature productivity, specific high energetic, and radioactivity could be achieved in autoclave by thermocouple.

7）直接引导问句：

You will find what want to find, said Professor Wang.

Professor Wang said, You will find what want to find.

8）文章中出现日期时，需要用逗号隔开日和年，如果日期按照月、日、年的顺序书写，日和年之间通常需要加逗号，若以这种方式出现在句子中，则年后也需要加逗号。如：

December 20, 2020

The international conference will start in ShenYang on December 20, 2019, and end on the next week.

如果书写的顺序为日、月、年则不需要加逗号，如：

20 December 2019

9）姓名和头衔之间使用逗号：

The professor who studies hydrometallurgy is W. Wang, Ph. D.

10）地址和地理名词之间使用：

His address is 200 Main street, Chaoyang District, Beijing.

11）数字中用逗号表示千位：

1, 024　　256, 000　　1, 564, 648　　123, 000

常见错误：

1）逗号后没有加空格；

2）英文书写中用顿号代替逗号。

（2）分号（semicolon）。英文写作中，分号的主要作用是连接和区分，使读者能够容易清楚的阅读。

1）区分未经连词连接但意义紧密相关的两个句子，两个句子可以是对比或因果的关系：

It is getting early, she don’t need get off now.

2）区分用连接副词相连的句子。常见的连接副词有 for example，furthermore，in contrast，nevertheless，on the other hand 等。如：

You do this by embedding; for example, the sideway approach within the top-down approach and in turn, the bottom-up approach.

（3）冒号（colon）。

1）区分主语与被说明内容：

There are two areas: point A and point B.

2）将方程与例子区别开：

There has a relationship with standard entropy $\overline{S^{\ominus}_{298(\mathrm{i,std.})}}$ as follows:

$$\overline{S^{\ominus}_{298(\mathrm{i,\ abs.})}} = \overline{S^{\ominus}_{298(\mathrm{i,\ std.})}} - 20.92z$$

3）区分主标题和副标题：

His paper is Complex Environments: The evolutionary line between individuals and groups.

4）信件谓语后：

Professor Wang:

Dear Mr. Li

常见错误：冒号前加空格。对于文章中的冒号，冒号后一定要加空格。当冒号的含义为比例时，冒号不需要加空格。

（4）破折号（dash）。破折号常表示事项列举分承，各项之前用破折号。常用于句子结构中的转变。如句子中的插入短语，词语用一前一后两个破折号。如：

Web browsers confine JavaScript—which is ubiquitous in web pages and advertisements, and runs automatically—to a sandbox supposed to prevent it from collecting private information.

插入短语或词语谓语句子末尾时，用一个破折号即可。

News house, larger schools, more sheep, more pigs and chickens, more horses and donkeys—everywhere we saw signs of prosperity.

使用时注意区分破折号“—”和连字符“-”。

（5）括号（parentheses）。括号一般是指表示文章中的注释部分使用的符号。这种注释是夹在正文中间的夹注。其目的是为了让读者了解得更透彻。

Tom Lantos(Chairman) warned fellow committee members they had a sobering choice to make.

However, Jerry Grandey(CEO) put a positive spin on the results in his comments.

需要注意的是，括号内的内容与括号间无空格，而左括号的前面和右括号的右边需要加空格，如果括号与标点符号相连，则不需要加空格。

误：You'll be able to catch James( visitor) all summer long as he tours our fine country.

正：You'll be able to catch James( visitor) all summer long as he tours our fine country.

（6）引号（quotation mark）。美式英文和英式英文使用引号有些区别。美式英文使用双引号“ ”划分引言，而英式英文用单引号‘ ’划分引言。

1）引言之前的说明词后常用逗号，但如果引言很短则会省略逗号：

The teacher said, “He is tired to be a teacher. ”

He yelled “Help! ” and called a police.

2）用于区分文章的标题：

His paper was entitled “Silence Spring”.

（7）方括号（brackets）。

1）作为评论或插入文章的补充信息：

Mr. Watson said, “I admire his [ Professor Bieber's ] behavior. ”

2）数学方程式中的括号：

$$\overline{c_p}(T) \approx \frac{[a_T + (b_T - 1)]\overline{S^{\ominus}_{298(i,\ \mathrm{abs.})}}}{\ln\left(\frac{T}{298}\right)}$$

3）参考文献的编号：

The heat capacity of pure substance can be obtained from the literature [12-15].

（8）省略号（ellipsis points）。

1）英文书写中省略号以三个中间有空格的句号显示，主要用处是引用资料的省略。在英文科技论文写作中，当我们需要引用其他作者的原文时，由于篇幅问题又想省略部分文字，则可以使用省略号。省略号表示被省略的部分。需要注意的是，如果省略号位于一句话的末尾，则需要使用四个带空格的句号。其中第一个句号表示句子的结束，后面三个句号表示省略。如：

The selection...dependent on the characteristics of the different waste streams, the total mercury concentration in waste is an important factor, but it would be insufficient to justify a recovery method...In most cases, speciation and coordination of mercury in waste is critical the choice of a recovery technology.

2）在方程或公式中，省略号用于表示一系列项目。如：

$$C_i,\ \text{where } i = 1,\ 2,\ 3,\ \ldots,\ n$$

$$F(t_1) = F(t_2) = \ldots = F(t_n)$$

（9）句号（period）。常用于表述一句话的结束，用于句子末尾。需要注意在英文中句号用“ . ”表示，而非“ 。”。以下列出几种句号的常用用法：

1）表示句子的结束：

陈述句：Something in my heart is broken.

祈使句：Let us measure the height of wall.

2）名字的缩写字母即缩写字的小写字母后：

L. G. Wang　　H. K. Li　　Dr.　　etc.　　et al.

3）若句子句尾为缩写字母则无须再加句号：

误：The winner of that game is James Bond, P. E. .

正：The winner of that game is James Bond, P. E.

4）句号可标在字母、罗马数字或阿拉伯数字后：

1. Abstract

2. Introduction

3. Method & Material

4. Result & Conclusions

5）在美式用法中，句号需要位于引号内：

His paper is entitled "A critical review on solvent extraction of rare earths from aqueous solutions."

6）词中间句号后不需要加空格：

e. g.　　i. e.　　a. m.　　p. m.

需要注意的是句子句号前不需要加空格。

（10）问号（question mark）

1）英文写作中问号用于直接疑问句的结尾。如：

What device can be used?

What is your paper title?

2）当问号和引号同时使用时，需要注意问号的位置，如：

Could you explain to me why this study method is called "high-temperature electrochemical"?

The policeman asked, "Where is your hometown?"

3）问号后需要加一个空格。

（11）感叹号（exclamation point）。

1）用于感叹句或表示感叹的陈述句后：

Help me!　　Hold your fire!　　Cover me!

2）不能用一个以上的感叹号表示惊叹。如：

误：Help me!!!

正：Help me!

## 4.3 前缀与后缀

通常将加在原单词前面的词缀称为前缀，而加在原单词后面的词缀称为后缀。一个单词加上前缀或后缀后，构成了含义与原单词相近或相反的新词，这种构词法叫派生法。通过单词的前缀或后缀，可以猜测一些单词的含义，本节将列举一些科技论文写作中常见的前缀与后缀作为参考，见表4-1～表4-2。

**表 4-1 常见的前缀**

| 前缀 | 中文释义 | 前缀 | 中文释义 |
|---|---|---|---|
| a- | 无，非，不 | ab- | 离开，相反，不 |
| acentric | 无中心的 | abaxial | 离开轴心的 |
| atypical | 非典型的 | abnormal | 异常的 |
| aperiodic | 非周期的 | absorb | 吸收 |
| ante- | 前，先 | anti- | 反对，相反，防止 |
| antecedent | 前项 | antiaging | 防衰老的 |
| antechamber | 预燃室 | anticlockwise | 逆时针的 |
| antevert | 前倾 | antimissile | 反导弹的 |
| by- | 次要的，附带的，副的 | co- | 共同 |
| by-channel | 支渠 | cochannel | 同频道的 |
| by-effect | 副作用 | coenergy | 同能量 |
| bypath | 侧管 | | |
| counter- | 反对，反抗，逆，对，交互，重复，副 | de- | 否定，非，相反 |
| countershaft | 副轴 | decomposition | 分解 |
| deca- | 十 | deci- | 十分之一 |
| decameter | 十米 | decigram | 分克 |
| decaploid | 十倍体 | decimal | 十进制的 |
| di- | 二；双 | dia- | 横过，通，全 |
| dicarbide | 二碳化物 | diagonal | 对角线 |
| dichloroethane | 二氯乙烷 | diameter | 直径 |
| dis- | 不，无，相反 | ef- | 出，离去 |
| disconformity | 不一致，不相称 | efflation | 吹出 |
| | | effluence | 流出 |
| ennea- | 九 | ex- | 出，外 |
| enneagon | 九角形 | exclude | 排外 |
| enneahedron | 九面体 | extract | 抽出 |
| extra- | 以外，超过 | hecto- | 百 |
| extrados | 外拱线 | hectogram | 一百克 |
| extra-low | 超低的 | hectowatt | 一百瓦 |
| extraneous | 外部的 | | |
| hemi- | 半 | hepta- | 七 |
| hemicycle | 半圆形 | heptagon | 七角形 |
| hemisphere | 半球 | heptahedron | 七面体 |
| hydro- | 水 | hexa- | 六 |
| hydrometallurgy | 湿法冶金 | hexagon | 六角形 |
| | | hexangular | 有六角的 |

续表 4-1

| 前缀 | 中文释义 | 前缀 | 中文释义 |
|---|---|---|---|
| homo- | 同 | hyper- | 超过，过多 |
| homocentric | 同中心的 | hyperbar | 超高压 |
| homothermic | 同温的 | | |
| infra- | 下，低 | inter- | 互相 |
| infrasonic | 低于声频的 | intercontinental | 洲际的 |
| infrared | 红外线 | international | 国际的 |
| iso- | 等；同 | macro- | 大，宏，长 |
| isoelectronic | 等电子的 | macmphysics | 宏观物理学 |
| isogon | 等角多角形 | | |
| isotope | 同位素 | | |
| meta- | 超 | micro- | 微 |
| metachemistry | 超级化学 | microwave | 微波 |
| milli- | 千分之一，毫，千 | mini- | 小 |
| milligram | 毫克 | minimum | 最小值 |
| millilitre | 毫升 | miniwatt | 小功率 |
| millimeter | 毫米 | | |
| mono- | 单一，独 | multi- | 多 |
| monatomic | 单原子的 | multi-roll | 多辊 |
| monocolor | 单色 | multishaft | 多轴 |
| monoxide | 一氧化物 | | |
| penta- | 五 | poly- | 多 |
| pentagon | 五角形 | polyatomic | 多原子的 |
| pentoxide | 五氧化物 | polycrystal | 多晶体 |
| pseudo- | 假 | quasi- | 类似，准，半 |
| pseudo-effect | 伪效应 | quasi-conductor | 半导体 |
| semi- | 半 | sept- | 七 |
| semiconductor | 半导体 | septangle | 七角形 |
| | | septuple | 七倍的 |
| sex- | 六 | sub- | 下，次，低 |
| sexangle | 六角形 | submetallic | 亚金属的 |
| sexivalence | 六价 | subcarbide | 低碳化物 |
| supra- | 超，上 | syn- | 共同，相同 |
| supraconductivity | 超导电性 | synchronism | 同步性 |
| | | synclastic | 同方向的 |
| tetra- | 四 | tri- | 三 |
| tetragon | 四角形 | triangle | 三角 |
| tetroxide | 四氧化物 | | |

表 4-2 常见的后缀

| 后缀 | 中文释义 | 后缀 | 中文释义 |
|---|---|---|---|
| -ane | 烷 | -ase | 酶 |
| butane | 丁烷 | lipase | 脂肪酶 |
| ethane | 乙烷 | pmtease | 蛋白酶 |
| methane | 甲烷 | | |
| -ene | 烯 | -hedron | ……面体 |
| butylene | 丁烯 | octahedron | 八面体 |
| ethylene | 乙烯 | polyhedron | 多面体 |
| propylene | 丙烯 | tetrahedron | 四面体 |
| -ide | 化物 | -meter | 计，仪，表，米 |
| copper oxide | 氧化铜 | barometer | 气压表 |
| potassium cyanide | 氰化钾 | hygrometer | 湿度计 |
| sodium chloride | 氯化钠 | kilometer | 千米 |
| -metry | 测量（学），度量（学） | -opia | 视力 |
| geometry | 几何学 | amblyopia | 弱视 |
| trigonometry | 三角学 | hyperopia | 远视 |
| | | myopia | 近视 |
| -valent | 价 | -yl | 基 |
| monovalent | 一价 | phenyl | 苯基 |
| divalent | 二价 | propyl | 丙基 |
| trivalent | 三价 | | |

# 4.4 大写与斜体

英文写作中大写字母、斜体字具有特殊的用法。本节将介绍大写字母和斜体的使用规则。

### 4.4.1 大写字母（capital）

（1）句子的句首字母一定大写。如：

It is well established from toxicology studies that mercury, both inorganic and organic, can cause serious health effects.

（2）如果句子在冒号后，作为独立分句，句首字母需要大写。如：

Determination of mercury in polluted soils surrounding a chlor-alkali plant: Direct speciation by X-ray absorption spectroscopy techniques and preliminary geochemical characterization of the area.

（3）专有名词必须大写：

1）人名、国家名、地名、语言名都需要大写。如：

Donald Trump Bill Clinton Justin Bieber
Chinese Turkish Germany
Beijing Hong Kong London

2）历史事件也需要大写：

World War Ⅱ the United Nations
the Environment Department the Civil War

3）日期和节假日等也需要大写：

November National Day

4）专有名词前的职称及专有名词后的学位等都需要大写：

President Bush W. Wang, Ph. D.

5）普通名词加于专有名词后形成专有名词时第一个字母需要大写，而普通名词为复数时，则不需要大写：

Central Street Yellow River
Silver Lake the Red Sea
Main streets South parks

（4）科学定律和原理及方法的专有名词必须大写，但其后的普通名词不需要大写。如：

Fick's first law Ohm's law

（5）文中指出同一本书中的标题或附录时，需要大写。如：
As shown in the Appendix.
The main finding was mentioned in the Result & Discussion.
（6）化学元素不需要大写。如：

carbon dioxide hydrofluoric acid

（7）图片、表格及公式、方程的名词通常需要大写。但有些期刊没有规定，具体还需作者参考期刊的格式规定。
（8）通常图表标题的第一个词和表格内第一个词的首字母都需要大写。但在一些期刊中要求图标和表标的每一词都大写，具体要求还需作者参考期刊的格式规定。

### 4.4.2 斜体字（italics）

（1）斜体字常用于书籍、期刊的题目或名称。如：
*Hydrometallurgy*
*Minerals Engineering*
（2）文中第一次出现的特殊词汇需要用斜体标注，而再次出现则不需要再用斜体标注。

（3）斜体还用于标注一些未成为英文词汇的外国词。如：

*Taijiquan*（太极拳）　　　　*Jiaozi*（饺子）

（4）在专业期刊中，通常都要求方程或公式中的变量用斜体标注。具体还需参照期刊的格式要求。

## 4.5　数字与计量单位

在科技论文写作中，数字和计量单位都有一般规则，作者写作时应当遵守，本节总结了一些作者应当遵守的数字和计量单位的一般规则。

### 4.5.1　数字

（1）科技论文中，小于十的整数通常用单词表示。如：

即用 one, two, three, … 代替 1，2，3，…。

（2）阿拉伯数字不能用作句首，必须写为英语单词。如果必须写数字，则应改写句子。如：

误：15 professors attended the international conference.

正：Fifteen professors attended the international conference.

（3）时间、计量值、小数和百分比等可以用阿拉伯数字表示。

（4）科技论文中小于十的序数词不能缩写，而大于十的序数词可以缩写。

| 正： | first | third | 31st |
|---|---|---|---|
| 误： | 1st | 3rd | |

（5）分数小于一时必须用单词表示。如：

One fourth of the apple　　　　One-half milliliter of water

（6）数字大于一百万时需要同时用阿拉伯数字和英文表示。如：

12 million　　　　26.52 million

### 4.5.2　计量单位

（1）英里、分、秒等测量单位缩写字母后需要加句号，其他缩量单位缩写字母后不需要加，例如：

| 15min. | 20sec. | 10mi. |
|---|---|---|
| 25km | 90kg | 6Hz |

（2）单位符号和数字之间需要空一个空格。如：

11km　　　　9kg　　　　7Hz

# 5 有色冶金论文投稿

## 5.1 期刊的选择

由于学科交叉广泛，研究手段丰富，有色金属冶金能够投的国际英文期刊并不仅仅限于冶金相关领域，还涉及电化学、材料科学、腐蚀科学等相关领域。以下列举一些常见的英文国际期刊。

（1）“Hydrometallurgy”。湿法冶金旨在汇编关于新工艺、工艺设计、化学、建模、控制、经济学和单元操作之间的接口的研究，为讨论事件分析和操作的难题提供平台。期刊主题包括：化学试剂或细菌作用在环境或升高的压力和温度下浸出金属值；从浸出液中分离固体；通过沉淀、离子交换、溶剂萃取、气体还原、胶结、电解精炼等方法去除杂质，回收金属值；通过焙烧或化学处理（如卤化或还原）对矿石进行预处理；试剂回收及废水处理。

（2）“Minerals Engineering”。该杂志主要目的是介绍有关矿物加工和提取冶金领域的最新发展。其广泛的研究和实践（操作）主题包括物理分离方法（如粉碎，浮选浓度和脱水）；化学方法（如生物、水、和电冶金）、分析技术、过程控制、模拟和仪器、矿物学方面的加工环境问题以及与可持续发展有关的环境问题。

（3）“Acta Materialia”。“Acta Materialia”提供了一个发表完整的原创论文和委托综述的平台，以促进对无机材料的加工、结构和性能之间关系的深入理解。寻求具有高影响潜力和/或在该领域有重大进展的论文。该期刊的结构包括原子和分子排列、化学和电子结构以及微观结构。重点是在机械或功能行为的无机固体在所有长度尺度下的纳米结构。其研究主题主要包括：尖端实验和理论的理解与属性、说明机制的合成和加工的材料特别与属性的理解、表征材料的结构和化学相关的具体属性的理解。同时，期刊欢迎采用理论和/或模拟（或数值方法）的论文，但这些论文应该通过与实验结果（在文献或当前研究中）进行比较，做出可测试的微观结构或性质预测，或阐明一个重要的现象，来证明与材料界的相关性。不鼓励主要集中于模型参数研究、方法开发或利用现有软件包获得标准或增量结果的论文。

（4）“Journal of Materials Science & Technology”。该杂志旨在加强材料科学与技术领域的科学活动的国际交流。期刊研究主题包括：金属材料、无机非金属材料复合材料等。

（5）“Journal of Alloys and Compounds”。“Journal of Alloys and Compounds”是一份国际同行评审的出版物，其内容包括化合物和合金材料。它的强大之处在于它所包含的学科的多样性，汇集了材料科学、物理冶金学、固体化学和物理学的成果。该杂志的跨学科性质很明显。材料问题的实验方法和理论方法需要各种传统的和新颖的科学学科之间相互作用。杂志研究主题不考虑液体合金、传统钢材、磨损、蠕变、焊接和连接、有机材料和聚合物、配位化学、离子液体、催化和生物化学等，此外，期刊不接受没有充分实验验证的纯计算论文。发表在期刊上的工作应包括对合成和结构的研究，以及对合金和化合物的化学和物理性质的调查，从而促进当前科学领域的发展。期刊要求提交发表的论文应包含新的实验或理论结果及其解释。

（6）“Corrosion Science”。“Corrosion Science”是一个为腐蚀领域提供交流的平台。其研究主题包括金属腐蚀和非金属腐蚀。这本国际期刊的范围很广。发表的论文从理论高度到实践高度不等，涵盖了高温氧化、钝化、阳极氧化、生化腐蚀、应力腐蚀开裂、腐蚀控制机理和方法等领域。

## 5.2 语言与格式问题

目前绝大多数的期刊都会将语言和格式要求列在网站上，可以在官网上查看。下面以期刊“Minerals Engineering”为例。大多数期刊都会要求使用包容性语言（use of inclusive language）。具体原因是包容性语言具有多样性，能够传达对多有人的尊重，促进机会平等。文章不该对任何读者的信仰做任何假设，不该包括任何可能暗示某个人在种族、性别、文化或任何其他特征上优于另一个人的内容，并且应该从头至尾使用包容性的语言。作者应该确保写作没有偏见。例如用 he/she 或 his/her 取代 he/his；用 chairperson 取代 chairman 等。

不同期刊对格式的要求各有差异，常见的要求如下：

（1）编号。通过编号将文章分成具有编号的明确的部分。比如小节编号为 1.1（然后是 1.1.1、1.1.2、…）、1.2 等，每个标题都应该显示在单独的行上。

（2）字体类型。大多数期刊对字体的建议为：Arial，Times New Roman，Symbol 或 Courier 等。

（3）图片。对论文中的插图按照出现顺序进行编号。图片的格式常要求为 TIFF 或 JPG，通常当单个图片大小超过 10MB 时，必须单独提供图片文件，不同期刊的要求不尽相同，投稿前应注意。确保每张图片都有一个标题。尽量减少插图中的文字，但要解释所用的所有符号和缩写。

（4）表格。提交的表格必须为可编辑文本，而不是以图像的形式。表格可以放在文章的相关文本旁边，也可以放在文章最后单独的页面上。根据不同期刊有不同要求。按照表格在文章中出现的顺序编号，每张表格都要有标题。确保表中显示的数据不会重复本文其他部分描述的结果。

（5）参考文献的格式。不同期刊对参考文献的格式要求也不同，插入参考文件时，建议搭配 EndNote 操作，插入时选择不同类型的参考文献即可。

## 5.3 投稿过程

### 5.3.1 投稿前的注意事项

（1）第一作者与通信作者（corresponding author）的区别。通信作者常指课题的总负责人，承担课题的经费、设计、文章的书写和把关。从知识产权的角度看，研究成果是通信作者的。而第一作者仅仅代表是课题最主要的参与者之一。

（2）挑选审稿人。很多 SCI 杂志需要自己提交与你研究成果相关的审稿人，常见的是 3 人，也有 5~8 人的。选择审稿人的主要途径如下：

1）询问自己的导师；

2）利用 SCI 或 SSCI 等检索和自己研究相关的科研人员；

3）相关期刊的编委和学术会议的主席、委员；

4）文章中参考文献的作者。

（3）重视审稿人的意见。审稿人的常见问题如下：

1）内容是否重要；

2）表述是否清晰；

3）证据是否充分；

4）参考文件的引用是否合理；

5）实验描述是否具有可重复性；

6）实验数据是否真实可靠；

7）图表使用是否规范。

回复审稿人意见时，一定要逐条回答，意见中要求补充的试验尽量满足，满足不了的也不要回避，需要给出合适的理由，此外，如果审稿人推荐了文献一定要引用。

（4）投稿前需要准备的材料。manuscript（初稿），tables（表格），figures（图片），cover letter（投稿信）。有时还有 title page（标题页），copyright agreement（版权协定），conflicts of interest（利益冲突）等，不同的期刊有不同的要求。

### 5.3.2 正式投稿

投稿过程大致可分为三步，首先需要提交论文，其次是编辑收稿，审稿人审稿，最后是期刊接收。

（1）提交论文。如今绝大多数期刊投稿均在互联网上进行，因此本篇仅介绍网上投递期刊的方式。登陆 ELSEVIER（https://www.elsevier.com/zh-cn），输入目标期刊名称进行搜索，以“Hydrometallurgy”为例，搜索结果如图 5-1 所示。

选择左边菜单“Guide for Authors”，显示页面如图 5-2 所示。

在“Guide for Authors”中，有期刊“Hydrometallurgy”对作者的投稿指导意见，其中主要包含投稿全部步骤。如从“Submission checklist”中可以看到需要上传的文件：

1）手稿（manuscript）。需要包含的内容主要有：标题页，脚注（有些期刊将此放在文本正文后）；摘要和关键字，引言，材料和方法（实验顺序），结果，讨论（或者结果和讨论连在一起），致谢，参考文献，所有数字（包括相关说明），所有表格（包括标题、说明、脚注）。

需要注意的是，在打印任何图形时，请明确指出是否应使用颜色。确保文本中的所有图形和表引用与提供的文件匹配。通常而言，每一部分开始于一张新页，每一张表格和图片使用一张单独页。所有页都应该在页底中部连续编号，和标题页连在一起。

2）图形摘要/亮点文件（如适用）。

Home > Journals > Hydrometallurgy

ISSN: 0304-386X

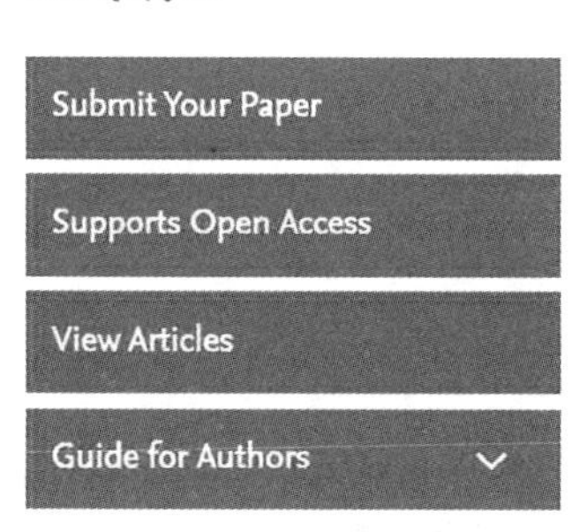




图 5-1 “Hydrometallurgy”搜索结果

图 5-2 “Guide for Authors”显示页面

3）补充文件（如适用）。

4）除此以外，还需要注意：手稿已被拼写检查和语法检查；参考文献列表中提到的所有参考文献都在文本中引用，反之亦然；使用其他来源（包括互联网）的版权材料已获得许可；提供竞争利益声明，即使作者没有竞争利益也需要声明；已详细阅读指南中期刊政策；根据期刊要求，提供推荐人建议和联系方式。

除了投稿前需要准备的材料之外，“Guide for Authors”中还有对文章结构的建议，图片、表格和参考文献等详细的格式要求，投稿前务必对照检查一遍。

在期刊的首页，左侧的菜单还有一栏为“Submit your paper”，即提交你的论文。点击后进入如下界面，如图 5-3 所示。

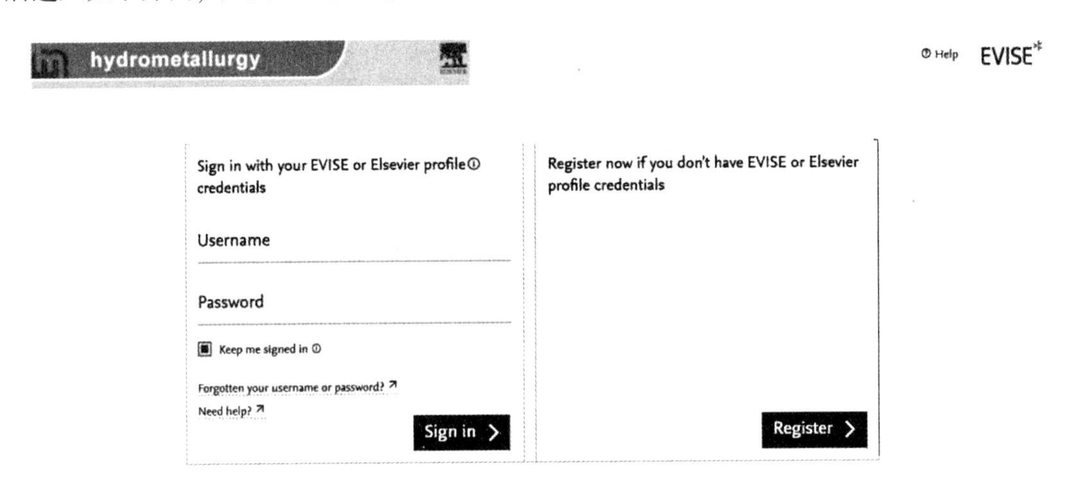


图 5-3　提交界面

注册账号，登录后按照提示依次完成 Select Article Type，Enter Title，Add/Edit/Remove Authors，Submit Abstract，Enter Keywords，Select Classifications，Enter Comments，Request Editor，Attach Files，最后查看 pdf 是否无误，确认无误后上传提交即可。

（2）修改论文。投稿后，如果没有选择编辑，则由期刊主编分派给其他编辑。编辑处理后一般会有两种状态。其一是“Decision letter being prepared,”这种情况是编辑未将论文给审稿人，而是自己做了决定。这种情况的原因可能是因为论文的语言不好或内容不好；第二种情况是“reviewer invited”，这种情况的意思是编辑已经将你的论文发给了审稿人。审稿的过程通常比较漫长。审稿后，审稿人的意见会返回给编辑，编辑会根据审稿人的意见做出接收还是修改的决定。如果需要修改论文，则需要根据审稿人的意见进行修改，可能需要补充试验或修改文章内容。当然，还可以选择进行申辩，申辩时注意谦虚谨慎的语气。通常大修后还是可能被拒，但小修大多数会被接收的。修改后再重新提交，重复以上步骤。

（3）接收。修改后的论文会再次被编辑发给审稿人审阅，最后各个审稿人对论文的评价都很好，则论文会很快被接收。如果论文被拒，就只能改投其他杂志了。

# 6 有色冶金相关国际会议

## 6.1 参会注意事项

国际学术会议本质上是学术交流活动，主要有以下几个目的：(1) 与国际同行建立联系；(2) 了解学科最新动向；(3) 分享研究成果；(4) 开拓研究思路。一般学术会议可以分为三类：首先是 conference，conference 主要指的是大型会议，谈论的主题、内容较多，范围广。其次是 workshop，指的是专题研讨会，其讨论范围较窄，研究主题针对性强，同时强调专业技术方面的交流。最后是 forum，forum 是论坛，一般指的是区域性的小型会议，与会人员一般为专门从事该领域研究的专家。作为一名科研人员，参加国际会议的可能性很大，因此，下面介绍一些参加国际会议的注意事项：

(1) 提前了解所参与会议的情况。如会议的规模，参会人员的情况，会议的主题等。会议通知往往是了解的渠道之一。获取会议通知的方式有以下几种途径：互联网、相关领域的期刊或组委会发来的通知等。得到会议通知后，首先应该确定会议研讨内容是否与自己的研究方向相关。如果相关就要开始准备材料，寄出自己的文章或摘要。

(2) 提前阅读会议的摘要，选择自己感兴趣的报告。国际会议中常常会有多个主题，提前阅读会议的摘要可以有效避免时间的浪费。

(3) 准备好口头报告和海报。国际会议的口头报告常用英语，因此对报告人的口语有一定要求。参加会议前应对报告内容和可能遇到的提问做好充分准备。

(4) 及时与会务组沟通。提前熟悉会议的注册、住宿、旅游等相关事务的安排。

(5) 提前办理签证。由于签证的办理需要一定的时间，应提前办好，避免影响出国参与会议。

## 6.2 口头报告

口头报告（oral presentation）是一场会议的主要环节，其目的是为作者和其他参会者提供一个详细交流的机会。会议报告有严格的时间控制，因此报告人需要控制好时间。会议的听众多数情况下是同行，少数情况下是小同行或非同行，因此报告需要做到深入浅出，突出重点，让多数人听得懂。报告之前，需要认真准备演讲稿和可视文稿。讲稿的框架与文章撰写的框架大致相同，可视文稿常为 PowerPoint。通常口头报告包含以下几个部分：

(1) 自我介绍；

(2) 研究的主题介绍；

(3) 研究内容、研究对象及研究结果；

（4）数据的分析；

（5）研究结果及讨论；

（6）研究的适用范围及意义；

（7）研究的目的及研究展望；

（8）结束发言表达谢意。

一般的口头报告包含以上几个部分。口头报告需要避免过于晦涩难懂的表述。使用少数人知道的专业术语前应对其进行解释。报告时脉络要清楚，让听众对其信服。语言避免拖沓啰唆。报告结束后，会议有问答环节，听众会对报告中感兴趣的部分有针对性的提问。可视文稿的框架与讲稿的框架基本一致，其内容主要为文章重点、数据、图标或公式等不便口头表达的内容。可视文稿可以提醒报告人按提前编排好的内容讲述自己的研究内容，还可以帮助观众理解报告人所讲述的内容。可视文稿的制作需要注意以下三点：

（1）尽量每一页只讲述一个重点。如果一页内容过多会分散听众的注意力。

（2）页面排版不要过于密集。行数过多读起来容易串行，字体过小不易辨认，因此减少行数以及增加行间距和字体大小会获得更好的效果。

（3）致谢需要分成两个部分。首先感谢项目的资助方，其次感谢最项目有过帮助的任何人，如提供过技术帮助等。

## 6.3 展　　板

展板（poster）是作者与听众进行交流和信息交换的媒介。展板与口头报告的区别在于传播观点的方式。口头报告主要报告人通过口头报告表述自己的观点或发现，而展板是通过展示内容表现作者的观点或发现。展板需要展示出作者最重要、最想传达的信息。然而，这并不意味着展板展示内容时，作者可以完全消失。实际上，在展示展板时，作者时常会站在自己展板的旁边，为感兴趣的听众提供更详细的信息。

展板通常放置于一些活动的现场，如茶歇（coffee break）或展销场所（trade shows）等地点。

国际会议中，展板环节的目的是为不同参会人员提供一个便捷的信息交换平台。展板常包含以下内容：

（1）题目（title）部分，告诉听众你的展板关于什么内容；

（2）摘要（summary）部分，阐述你做了什么，如何做的，以及主要的发现结果，关键的结论，观点等；

（3）介绍（introduction）部分，这部分应包含你试图解决的问题；

（4）试验原理或方法（theory & methodology）部分，这部分应解释你所使用技术的原理或试验方法，除此以外你还应该表述你的假设及其合理性；

（5）试验结果（result）部分，本部分应包含试验的主要结果；

（6）结论（conclusion）部分，本部分应包含试验的主要结论；

（7）研究展望（further work）部分，除了以上内容外，还应包含本次工作如何进一步研究的想法或方法等。

以上是展板常见结构。除了内容外，展板的字体、颜色、字号甚至是线条等也需注意。优秀的展板可以迅速帮助听众找到展示内容的重点。展板的设计没有绝对的标准，不同人不同学科有不同的审美水平和需要，因此，以下提供几条通用原则作为参考：

（1）展板全文保持一致的风格。不一致的展板的风格会带给听众不和谐的感受，并且会影响信息表达的流畅性。展板使用的图片应尽量具有相同的尺寸。图片的标题应在图片的顶部或底部。

（2）选择合适的图片。展板中使用的图片的大小应清晰可见。如果选择使用线形图，那么应该用具有对比性的颜色表示不同类型数据，而不是不同的线条特征。

（3）不要全都使用大写字母。阅读全部为大写字母的展板会为听众带来很大困难，因此一定要避免。

（4）不要使用超过两种类型的字体。展板的字体类型过多会分散听众的注意力。选择字体的原则第一要容易阅读，第二需要在一定的距离外也能够看得清楚。常用字体推荐 Times New Roman 和 Arial。

（5）展板内容的选择应丰富但不冗杂。展板的每一处空白都应该被充分利用，但这不意味着需要把研究的全部内容全部塞进去，展板内容的选择一定要有的放矢，过多的信息会使听众找不到重点，并且造成视觉疲倦。

# 7 英文写作辅助工具

随着互联网科技的发展，一些公司开发了许多优秀的应用工具为学习英语写作的同学提供帮助。无论是词汇、短语，还是常用搭配，甚至语法，这些工具都能在短时间内帮助我们找到正确的英语表达。以下介绍一些笔者常用的英文写作辅助工具。

## 7.1 Linggle

Linggle 搜索引擎是一款可以查询英语语法和词汇搭配的工具（官方网站 http：//linggle. com）。网站首页有使用方法的说明，笔者再次介绍几种常用的方法。

（1）查询常见词汇搭配。当你不确定使用的单词时，可以利用“_”代替需要插入的词汇，网站通过大量例句挑选出地道的常用搭配。

例如：不确定 present a method 后应该加什么介词，可以在网站中搜索 present a method _，网站将会出现与该句型搭配的常用介词以及其使用频率，如图 7-1 所示。

linggle10[12] present a method _

| N-gram | Percent | Count | Example |
|---|---|---|---|
| present a method for | 48.8 % | 12,227 | Show |
| present a method to | 22.8 % | 5,726 | Show |
| present a method of | 11.9 % | 2,983 | Show |
| present a method that | 7.9 % | 1,986 | Show |
| present a method which | 3 % | 758 | Show |
| present a method based | 2.1 % | 536 | Show |
| present a method by | 0.8 % | 198 | Show |

图 7-1 搜索界面

（2）检验搭配的正确性。当使用了不确定的词汇后，可以通过在词汇前加“?”查询词汇的使用是否正确。

例如：不确定 discuss about the issue 中介词 about 的使用是否正确，可以在网站中搜索 discuss ? about the issue ，结果如图 7-2 所示。

结果表明 discuss the issue 的使用频率远高于 discuss about the issue，因此写作时我们应该选择 discuss the issue 进行表述。

（3）查询更常见的表达。当遇到两个词不知道如何选择时，可以将两个词同时输入到搜索框中，并在两次中间加上“/”，如图 7-3 所示。

（4）查询词语搭配。当我们遇到某些名词，不知道用什么动词搭配的时候，可以在名词前加“v.”，这样就会搜索出与该名词搭配的动词。例如我想知道 investigation 应该与什么动词搭配，可以在搜索框输入 v. investigation，结果如图 7-4 所示。

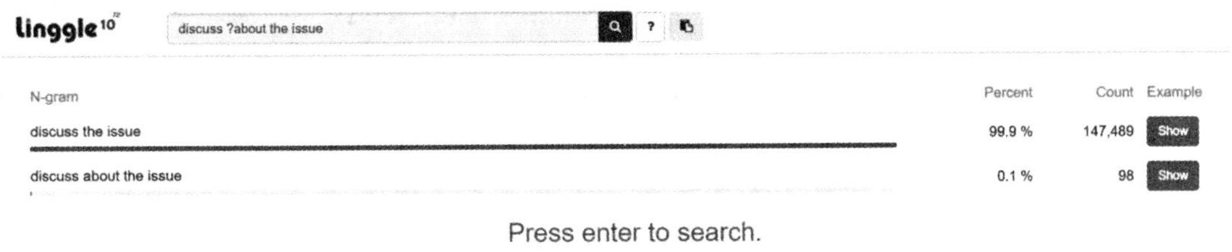

linggle10　discuss ?about the issue

| N-gram | Percent | Count | Example |
|---|---|---|---|
| discuss the issue | 99.9 % | 147,489 | Show |
| discuss about the issue | 0.1 % | 98 | Show |

Press enter to search.

图 7-2　搜索结果

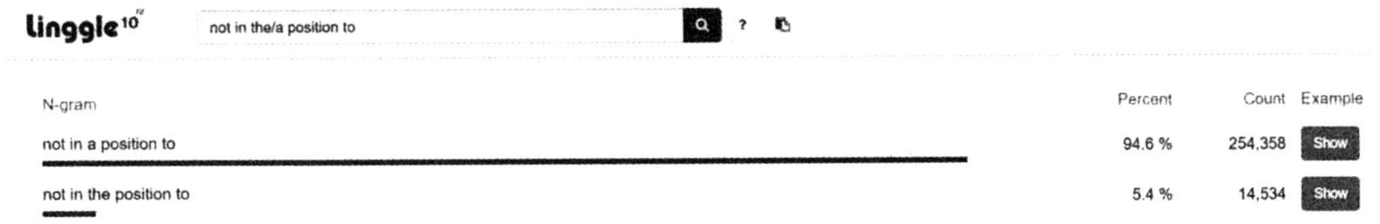

linggle10　not in the/a position to

| N-gram | Percent | Count | Example |
|---|---|---|---|
| not in a position to | 94.6 % | 254,358 | Show |
| not in the position to | 5.4 % | 14,534 | Show |

图 7-3　搜索结果

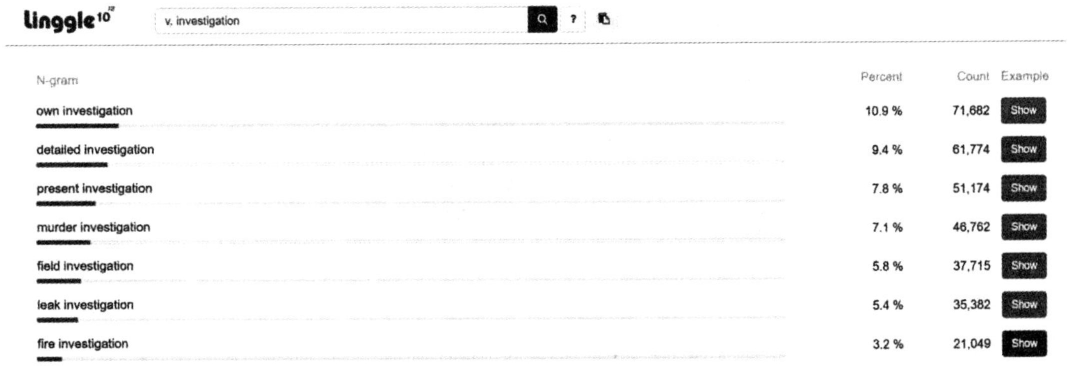

linggle10　v. investigation

| N-gram | Percent | Count | Example |
|---|---|---|---|
| own investigation | 10.9 % | 71,682 | Show |
| detailed investigation | 9.4 % | 61,774 | Show |
| present investigation | 7.8 % | 51,174 | Show |
| murder investigation | 7.1 % | 46,762 | Show |
| field investigation | 5.8 % | 37,715 | Show |
| leak investigation | 5.4 % | 35,382 | Show |
| fire investigation | 3.2 % | 21,049 | Show |

图 7-4　搜索结果

## 7.2　Grammarly

Grammarly 是一款功能强大的语法纠错软件，可以修改常见语法错误，地道表达（官方网站 https：//www. grammarly. com/）。目前提供了网页版、Mac 版和 Windows 版。官网还提供浏览器的扩展插件，此外 Windows 版还提供了 Word 插件，使用 Word 写作时非常方便。目前分为免费版和高阶版。免费版提供语法和拼写检查，而高阶版在此基础上还提供词汇使用建议及查重功能。

当文中出现错误时，画线部分为需要修改的地方，Grammarly 会提供详细的修改建议，如图 7-5 所示。

## 7.3　Semantic Domains

英文写作时，表达同一个意思时通常不会一个词一用到底，一方面是为了避免行文枯燥；另一方面是避免过高的重复率。Semantic Domains 正是满足这一需要的网站（官方网

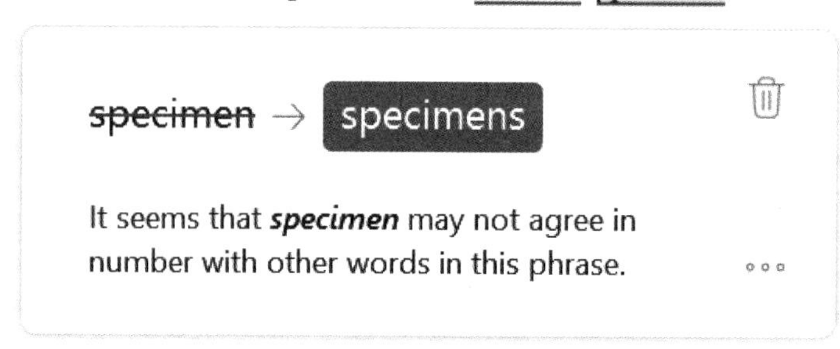


图 7-5 Grammarly 修改建议图

站 https：//www. semdom. org/），当你需要对某些词语进行替换时，可以在该网站上进行搜索，寻找替换词汇。例如需要找与 investigate 意思相近的词，可以在网站搜索 investigate，会出现如图 7-6 所示结果。

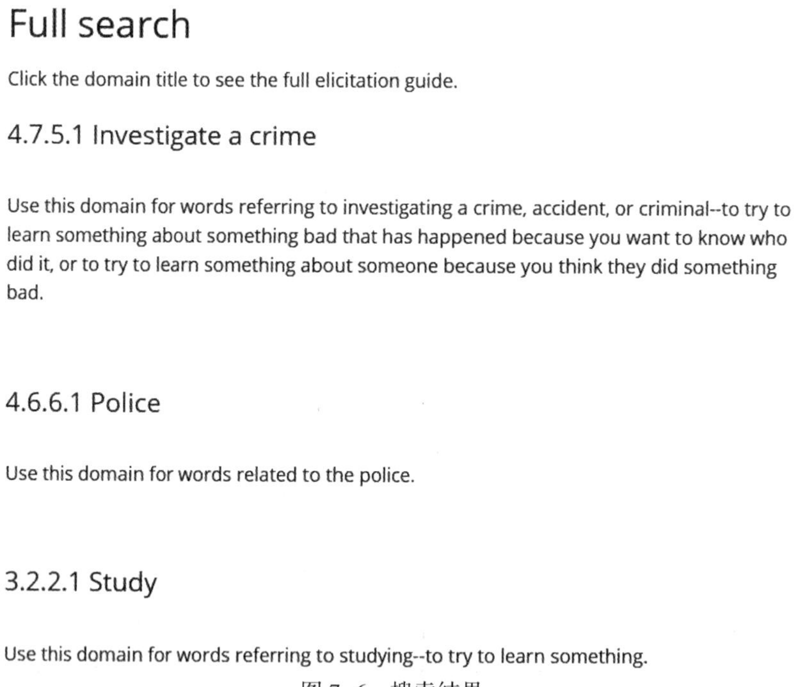

Full search

Click the domain title to see the full elicitation guide.

4.7.5.1 Investigate a crime

Use this domain for words referring to investigating a crime, accident, or criminal--to try to learn something about something bad that has happened because you want to know who did it, or to try to learn something about someone because you think they did something bad.

4.6.6.1 Police

Use this domain for words related to the police.

3.2.2.1 Study

Use this domain for words referring to studying--to try to learn something.

图 7-6 搜索结果

搜索结果列出了 investigate 的三个意思，其中第三个为我们所需要的，继续点击可以得到如图 7-7 所示结果。

网站又将 study 的含义进行细分，写作时我们挑选需要的词汇即可。

## 7.4 Academic Phrasebank

Academic Phrasebank 上收集了大量的规范的学术表达（官方网站 http：//www. phrasebank. manchester. ac. uk/），根据网站首页的菜单栏可以找到文章不同章节所需要的短语，内容十分丰富，如图 7-8 所示。

## 3.2.2.1 Study

Use this domain for words referring to studying--to try to learn something.

**OMC Codes:** 120 Research Methods
Methodology
121 Theoretical Orientation in Research and Its Results
122 Practical Preparations in Conducting Fieldwork
123 Observational Role in Research
124 Interviewing in Research
125 Tests and Schedules Administered in the Field
126 Recording and Collecting In the Field
127 Historical and Archival Research
128 Organization and Analysis of Results of Research
**Louw Nida Codes:** 27D Try To Learn

**What words refer to studying something?**

*study, analyze, check, concern yourself with, consider, determine, examine, experiment (v), follow, inquire into, investigate, learn, look into/for, look something up, probe, prowl, seek, search, research, study, test, trace, track, trail, try to find out about,*

**What words refer to the process of trying to learn something?**

*evaluation, examination, experiment (n), investigation, inquiry, quest, study, search*

**What words refer to the subject that is being studied?**

*subject, topic, question*

**What words refer to studying something a second time?**

*review, revise (British)*

**What words refer to a person who studies something?**

*astronomer, biologist, botanist, detective, investigator, linguist, mathematician, operative, reconnoiterer, scientist, scout, spy, undercover agent, zoologist*

**What words describe someone who likes to study?**

*studious, inquisitive,*

图 7-7　搜索结果

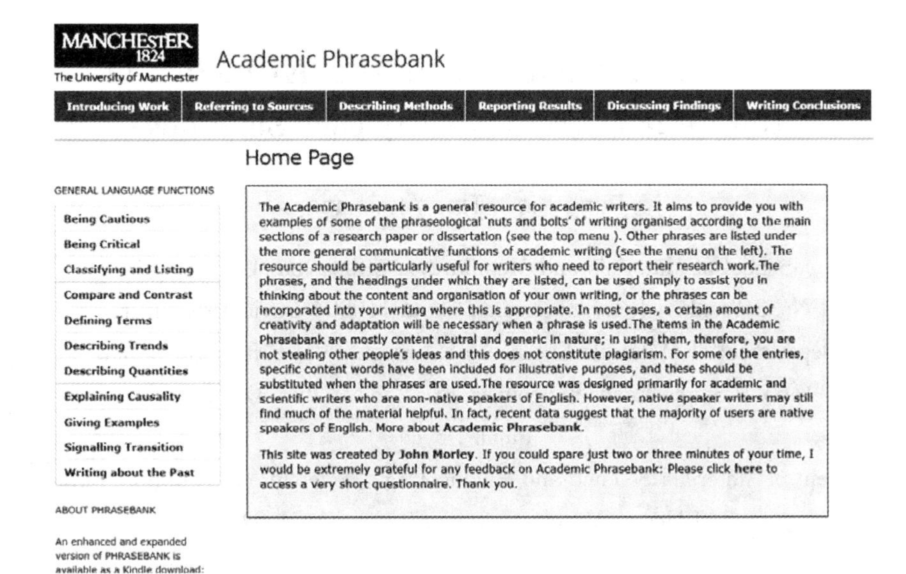


图 7-8　搜索结果

以上是笔者常用的几款英文写作的工具，安装及操作都很方便。除了以上介绍的几款英文写作辅助工具外，还有诸如 Ludwig，Netspeak，Justtheword 等辅助软件，功能与上述提到的软件类似，在此不予赘述。英文写作的提高主要还是依靠英文水平的增长，多读、多写才是最为实用的提高写作能力的方法。

# 8 有色冶金期刊论文范例

本书选编了三篇已发表的有色金属冶金领域的期刊论文，包括两篇综述性论文和一篇技术论文。这些科技论文语言规范，难度适中。范文保留了论文的草稿和由专业人士修改的记录，具有一定的参考价值。需要指出的是，以下只是这几篇论文接近完稿发表时的部分修改记录，仅供读者论文写作时参考，论文最终在期刊的发表版与此可能有不同之处，请参阅论文各自期刊。

本章所有范例中括号内的灰色部分为删除内容，下划线部分为修改内容。

## 8.1 范例一

本篇论文堪称本领域的一篇经典论文，论文从初始创作到最终投递至“Minerals Engineering”期刊发表，作者曾修改论文草稿 10 余次，并经数位本领域著名学者专家审阅。论文自发表后（2014 年），获得了本领域专家学者的一致好评，至今已连续数年为 ESI 高被引论文及该期刊的“most cited”和“most downloaded”论文。

### A C(*c*)ritical Review on Solvent Extraction of Rare Earths from A (*a*) queous Solutions

Feng Xie*, Ting An Zhang

School of Materials and Metallurgy, Northeastern University,
No. 3 Wenhua Road, Shenyang, China 110004 David Dreisinger
Department of Materials Engineering, University of British Columbia,
309-6350 Stores Road, Vancouver, BC, Canada V6T 1Z4 Fiona Doyle
Fiona Doyce
Department of Materials Science and Engineering, University of California, Berkeley
210 Hearst Mining Building, Berkeley, CA 94720

**Abstract**

R(*r*) are earth elements (*are uniquely and*) have unique physicochemical properties that make them critical in many high-tech industries(*y nowadays*). Bastnesite (La, Ce) $FCO_3$, monazite, (Ce, La,

* Corresponding author.
Feng Xie: Tel. : +86 24 8368-7729; E-mail address: xief@ smm.neu.edu.cn

Y, Th) $PO_4$, and xenotime, $YPO_4$, are (*commercially*) the most important commercial (*re*) sources of rare earths. (*In practice,*) R (*r*) are earth minerals are usually (*concentrated*) beneficiated by flotation or gravity methods to produce concentrates that (*is*) are subsequently leached with (*an*) aqueous inorganic acids, such as HCl, $H_2SO_4$, or $HNO_3$. After solution purification, (*separation processes based on*) solvent extraction (*techniques are*) is usually used to (*yield*) separate individual rare earths or produce mixed rare earth solutions or compounds. Rare earth producers follow (*almost identical*) similar principles (*or*) and schemes when (*on*) selecting (*process routes for*) specific solvent extraction routes(*separation of rare earths from an aqueous solution*). The use of cation exchangers, solvation extractants, and anion exchangers, for separating rare earths has been extensively studied. The choice of extractants and aqueous solutions is influenced by both cost considerations and requirements of technical performance. Commercially, D2EHPA, HEHEHP, Versatic 10, TBP, and Aliquat 336 have been widely used in rare earth solvent extraction processes. (*On the aspect of equipment, u*) Up to hundreds of stages of mixers and settlers may be assembled together to (*provide mixing and settling*) achieve the necessary separations. (*Separation*) This paper reviews the chemistry of different solvent extractants (*chemistry*) and typical configurations for rare earth (*solvent extraction processes have been reviewed in the paper*) separations.

**Keywords:** Rare Earths; Solvent Extraction; Process Configuration

(文章目录略)

## 1 Introduction

### 1.1 Rare earths ore

The term rare earths was originally used to designate the oxides of scandium, yttrium, lanthanum and the 14 elements following lanthanum in the periodic table of elements, i. e. from cerium to lutetium inclusive. (*Later*) More recently, the term "rare earths" has been used to designate the elements themselves. Scandium and yttrium (*are considered rare earth elements since they*) tend to occur in the same ore deposits as the lanthanoids and exhibit similar chemical properties. The term "rare" earth is a misnomer; (*Rare earths*) they are relatively abundant in the Earth's crust, however, (*because of their geochemical properties, rare earth elements*) they are typically dispersed and (*not often found*) only rarely occur in concentrated and economically exploitable (*rare earth*) mineral deposits(*, leading to the term "rare earth"*).

Rare earth mines (*emerged* ) operated in South Africa, India, and Brazil in the 1950s, but from the 1960s to the 1980s, the (*greatest*) largest global producer was a mine in Mountain Pass, California. In the 1990s, China began the large scale (*exploitation*) production and export of cheaper rare earths (*with a much cheaper price*). Other producers were unable to compete economically, and (*Most other countries stopped producing rare earths due to this financial pressure and mines around the world*) began closing in the 1990s, with the Mountain Pass mine shutting down in 2002. Consequently, China is currently the world's largest producer of rare ear elements by an extreme margin, providing more than 95% of the world's total supply from its mines in Inner Mongolia. The world reserves and production of rare earths are summarized in Table 1.

Though av (*v*) ariety (*of*) rare earth minerals are known, notably(*have been identified,* ) bastnasite (La, Ce) $FCO_3$, monazite, ( Ce, La, Y, Th) $PO_4$, and xenotime, $YPO_4$ (*, are the commercially important resources of rare earths*). Bastnasite deposits in China and the United States constitute the largest percentage of the world's rare earth(*economic*) resources. Notable occurrences include t(*T*) he (*huge*) carbonatite-hosted bastnasite deposit (*was discovered*) at Mountain Pass, California, several (*while Chinese deposits of*) bastnasite deposits(*were found several*) in Sichuan Province, China, and the massive deposit at Bayan Obo, Inner Mongolia, China. Monazite deposits in Australia, Brazil, China, India, Malaysia, South Africa, Sri Lanka, Thailand, and the United States constitute the second largest segment. Apatite, cheralite, eudialyte, loparite, phosphorites, secondary monazite, spent uranium solutions, and xenotime make up most of the remaining resources. A very large resource enriched in heavy rare-earth elements is inferred for phosphorites of the Florida Phosphate District. A special type of rare earths resource, the ion-adsorption rare earth deposits (rare-earth-bearing clays) are (*widely distributed in southern China*). In ion-adsorption rare earth(*these*) deposits, widely distributed in southern China, (*rare earths occur in the crust of*) weathering (*mineral deposit and are mainly*) has left rare earths adsorbed on the surface of clay minerals such as kaolin, feldspar, and mica, etc.

### 1.2 (*Rare*) Technological applications of rare earth (*application*)s

(*Rare earth metals and their compounds are used in, and indeed are often crucial for, a broad and rapidly expanding range of applications that rely upon their The diverse*) Chemical, catalytic, electrical, magnetic, and optical properties (*of rare earth elements (REE) have led to a rapidly increasing variety of applications involving these metals*). rare earths are widely used in (*the*) traditional sectors including metallurgy, petroleum, textiles, and agriculture. (*They*) As indicated in Table 2, they are also becoming uniquely indispensable and critical in many high-tech industry such as hybrid cars, wind turbines, and compact fluorescent lights, flat screen televisions, mobile phones, disc drives, and defense technologies. (*A partial listing of critical rare earth applications is shown in Table 2. Depending on the requirement of applications, d*) Different rare earths are needed to (*enable*) supply the required functionality in these applications(*of the applied materials*). In some cases, a single rare earth element may be required, such as La for nickel-metal hydride batteries, but other applications require(*; in other cases,* ) a mixture of rare earths (*can be applied to realize the required functions*), for(*Typical*) example(*s are*) Nd and Pr for rare earth magnets and Eu (or Tb) and Y for rare earth phosphors.

Rare earth-containing permanent magnets are (*made from*) alloys of rare earth elements and transition metals such as iron, nickel, and cobalt. (*There are two types of rare earth magnets, neodymium magnets and samarium-cobalt magnets. The*) S(*s*) amarium-cobalt magnets were first developed in the 1970s. However, due to their higher cost and weaker magnetic field strength, these magnets are now used less (*used*) than neodymium magnets, unless their higher Curie point is needed (*nowadays*). Neodymium permanent magnets, (*which are made from*) a(*n*) tetragonal alloy of neodymium, iron, and boron *(to form the)* ( $Nd_2Fe_{14}B$) *(tetragonal crystalline structure)*, have been used in a wide range of applications requiring a high energy product and high coercive force *(application)*. *(Previous work has demonstrated that)* N*(n)* eodymium can be replaced by praseodymium and up to 5% cerium in high energy product magnets. *(It was also reported that t)* The addition of terbium and dysprosium can enhance the coercivity of Nd-Fe-B sintered magnets.

*(Rare earths phosphors are widely used in)* High efficiency lighting, flat display screens, plasma screens, and liquid crystal screens due to their unique luminescent properties . *(The unique property of rare earth phosphors originates from the fact that)* Unlike transition metal ions, the spectral position of the emission lines of rare earths is almost independent of the host lattice(*, in contrast to line emission generated by transition metal ions*). *(Some of r)* Rare earth ions (such as $Tb^+$ and $Eu^+$) emit at *(specific positions, )* frequencies that enable*(ing)* high lumen efficacies and a good quality of white light *(which is adapted to the human eye in a comfortable way, )*. *(When)* Replacement of some of the rare earth cations *(in)* of a *(pure)* crystalline rare earth phosphor *(are replaced)* by ions of another rare earth element *(an impurity or activating substance)* activates the phosphor, achieving a high degree of fluorescence *(is achieved)*. For example, terbium-activated gadolinium oxysulphide ( $Gd_2O_2S$: Tb) gives a maximum fluorescence when about 0. 3% of the gadolinium atoms have been replaced by terbium. *(Yttrium)* Commercial screens have used yttrium tantalates activated by thulium ( $YTaO_4$: Tm) or niobium ( $YTaO_4$: Nb) *(are also used in commercial screens)*.

### 1.3 Primary rare earth extraction process

As described above, bastnasite, monazite, and xenotime are the principal rare earth minerals of commercial importance. Typical compositions for these minerals are shown in Table 3 ( there may be significant compositional variations depending on sources). *(Though v)* Various processing routes have been developed to*(for)* recover*(ing)* rare earths. After mining and comminution *(from different ores have been developed)*, *(rare earth minerals are usually concentrated)* ore is beneficiated by flotation or gravity methods to produce rare earth concentrates, which then undergo *(that can be fed to subsequent)* hydrometallurgical processing to recover rare earth metals or compounds.

The existing process for treating simple bastnasite concentrates *(is quite)* are relatively straightforward to treat. In order to reduce the acid consumption, bastnasite concentrates are usually roasted to decompose carbonate before *(under)* leaching with either hydrochloric *(acid)* or sulfuric acid. It is desirable for leaching to be as selective as possible, to minimize the processing costs *(for)* associated with separating undesired elements from the leach solution. Separation costs can also be lowered by *(first)* removing cerium from *(the stream of)* solutions containing the other lanthanides; *(, )* since cerium *(is the major component)* comprises about half of the rare earths in *(of)* bastnasite, removing it gives a commensurate reduction in the capacity needed during solvent extraction *(and monazite)*. *(This can be accomplished by oxidizing c)* Cerium is easily removed by oxidizing to $CeO_2$ during roasting. $CeO_2$ does not dissolve readily in acidic lixiviants, and reports principally to the leach residue, from which it can then be recovered separately. However, care is needed to provide an adequately oxidizing environment at the high temperatures, to ensure complete oxidation to

$CeO_2$. Hydrochloric acid may also promote (*favors*) reduction of Ce(Ⅳ) and hence, incomplete separation of Ce(Ⅲ) from the other trivalent lanthanides, Ln(Ⅲ). Other processing routes (*may be designed to*) leach all the rare earths, then oxidize cerium in the aqueous phase before precipitating it and separating by filtration. For example, $Ce(OH)_4$ was precipitated at the T(*t*) horium plant using ammonium carbonate and ammonium persulfate. In some cases, sodium hypochlorite is also used to oxidize dissolved Ce(Ⅲ).

In (*the principle*) Baotou (*Rare Earths Plant*), the largest producer of rare earths in China, the bastnesite concentrates contain a small amount of monazite. Figure 1 shows a typical leaching process used for Baotou rare earth concentrates, which is flexible and can accommodate different concentrates. The (*principal*) process (*for treating these concentrates*) starts with roasting with concentrated sulfuric acid at (*an elevated temperature*(>300℃); *this is needed*) to "crack (*break-down*)" the monazite. (*One typical leaching process of Baotou rare earth ore is shown in Figure 1. A significant advantage of the process is its flexibility in being able to process different concentrates. However,*) The rare earth sulfates formed during this process are then leached with water, and excess acid is neutralized with magnesia. The leach solution then proceeds to solvent extraction, or a mixed rare earth chloride (for electrolysis to misch metal) is produced by precipitation with ammonium carbonate, followed by dissolution with HCl and crystallization. Unfortunately, the radioactive element, thorium, is precipitated and reports to the leach residue. (*and*) It can not be recovered economically, resulting in both loss of the valuable thorium and (*severe*) potential environment hazards(*pollution*). HF and sulfur dioxide report to the off-gas from roasting. Large amounts of water or alkaline solutions are needed to remove them(*cover the HF and sulfur dioxide from the off-gas*), resulting in (*production of*) large volumes of (*waste*) acidic (*solution*) effluents.

Some modified The roasting process(*es*) has(*ve*) been modifi(*develop*) ed, (*e. g., roasting rare earth concentrates with*) for example by adding MgO or (*with*) CaO and NaCl to stabilize fluorine in the leaching residue (*rather tha*) instead of(*n*) releasing(*e*) it (*in*) to the waste gas phase. (*The use of*) Bastnasite has also been roasted with ammonium chloride, which decomposes into gaseous HCl that forms rare earth chlorides, which are readily leached with hot water(*roasting for extracting rare earths from bastnesite was also developed. In this process,* $NH_4Cl$ *decomposes into gaseous HCl under heating which reacts with the rare earth oxide to form rare earth chloride which can be easily leached with hot water.*) Another variant involves heating concentrate with sulfuric acid(*A process using a moderate temperature (150-250℃) to decompose the mixed concentrates was also developed, in which the concentrate was first mixed with sulfuric acid and heated*) at 40-180℃ for two to four hours before roasting at 150-330℃. Th(*e*) is suppresses decomposition of sulfuric acid (*is significantly suppressed*), resulting in a relatively high(*er*) fraction of HF in the gas phase; this(*, which*) can be recovered as solid $NH_4F$ by reacting with $(NH_4)_2CO_3$ in the off-gas pipe. In some plants, the bastnasite concentrates are first digested with concentrated NaOH to decompose carbonates and then leached with hydrochloric acid to produce mixed rare earth chlorides. The disadvantages of this process include high alkaline consumption and the radioactive thorium reporting to both the leachate and the residue, which hampers subsequent recovery.

Monazite and xenotime concentrates can be leached either by sulfuric acid or by sodium hydroxide at elevated temperature to decompose the orthophosphate lattice. The sodium hydroxide treatment is preferred in most commercial extraction plants because it better separates phosphate from the rare earth. The ion-adsorption type rare earth ores are usually leached directly with inorganic acid, either in dumps or in-situ, resulting in dissolution of most of rare earths in acidic solutions.

Regardless of the original rare earth mineral or the precise leaching process, the leach solution will usually contain dissolved impurities such as iron, which are removed by precipitation before proceeding to solvent extraction to separate the rare earths.

## 2 Solvent Extraction Separation of Rare Earths

(*Solvent extraction generally starts by separating As discussed above, rare earth concentrates or calcines are usually leached with an inorganic acid such as HCl for bastnesite and sulfuric acid for bastnesite calcine. After solution purification, separation processes are needed to separate* ) Different groups of rare earths from the leachate. While some primary rare earths producers may choose to sell intermediate, mixed products (*only*), others (*may*) perform different downstream separations (*in order*) to produce individual rare earth salts or oxides (*for use in applications such as yttrium in phosphors and neodymium in magnets*). I(*Separation of i*) ndividual rare earths are difficult to separate from each other, (*is usually difficult*) due to their similar(*ity in*) physical and chemical properties. S(*ome s*) eparation processes based on ion-exchange and solvent extraction techniques have thus been developed to produce(*yield*) high purity single rare earth solutions or compounds. Before the advent of industrial scale solvent extraction in the 1960s, ion exchange (*For many years, the resin* ) technology was the only practical way to separate the rare earths in large quantities (*before the implementation of the solvent extraction process on an industrial scale in 1960s*). Nowadays, ion exchange is only used to obtain small quantities of(*only small amounts of rare earth elements are separated by ion exchange resins for obtaining*) high purity rare earth product for (*such as*) electronics or analytical applications. (*The use of*) S (*s*) olvent extraction (*technique for separating and purifying rare earths from aqueous solutions*) is generally accepted as the most appropriate commercial technology for separating rare earths (*separation*).

S(*ome s*) olvent extraction processes for separation and purification of rare earths have been reviewed (*previously*) during the 1990's. Table 4 summarizes c(*C*) ommercial extractants reported in the literature for rare earth solvent extraction are summarized in Table 4. (*Three* ) All three (*different*) major classes of extractant(*s*), namely, cation exchangers (or acidic extractants), solvation extractants (or neutral extractants), and anion exchangers (or basic extractants), have been utilized (*commercially*) for separating rare earths. Some chelating extractants have also been suggested for rare earth separations.

### 2.1 Cation exchangers

The overall extraction of rare earth elements from aqueous media by cation exchange extractants in

their acidic form(*extractants*) can generally be expressed as: (公式略)
where Ln denotes any rare earth, A denotes the organic anion, and overscoring denotes species present in the organic phase. Equation 3 is somewhat simplified; the(*The*) acidic extractants are usually aggregated as dim(*m*)ers or larger oligomers in non-polar organic solutions, which (*to*) lowers their polarity, and the rare earth complexes formed upon extraction may contain undissociated organic acid(*, and sometimes even contain hydroxide ions*). Thus a more accurate depiction of the extraction reaction (*between a rare earth cation and an acidic extractant is usually expressed as*) is: (公式略)
Here $H_2A_2$ refers to the dimeric form of (*either carboxylic or organophosphorous*) the organic acid. From inspection of equations 3 or 4(1), it is evident that the extraction of rare earths with cation exchangers is promoted by increasing the aqueous phase pH, while the stripping process, which reverses the extraction reaction, is promoted by increasing the acidity of the aqueous stripping solution.

(*In literatures, two terms, the distribution coefficient and the separation factor, are commonly used for comparing the extractability of the extractant with the rare earth metal ions. They are defined as following:*) (公式略)
(*where D represents the distribution coefficient of a rare*) (*[M] denotes the molar concentration of a rare earth metal in the organic phase and [M] is that in the aqueous phase;*) (公式略)
(*where $\beta_{M1/M2}$ is separation factor of two metals, $M_1$ and $M_2$; $D_{M1}$ and $D_{M2}$ are the distribution coefficients of metals $M_1$ and $M_2$, respectively.*)

Two different classes of cation exchangers are use for rare earth separations, namely carboxylic or fatty acids, and organic derivatives of phosphorous acids (*and carboxylic or fatty acids*).

### 2.1.1 Carboxylic acids

The use of different carboxylic acids, including naphthenic acids and Versatic acids, for extracting rare earth metal ions has been reported. The extraction behavior of yttrium differs for these reagents; yttrium is extracted by Versatic 10 with the middle rare earths (La<Ce<Nd<Gd<Y<Ho<Yb) whereas it is extracted by naphthenic acid with the light rare earths (La<Ce<Y<Nd<Gd<Ho <Yb). Zheng et al. noted that the behavior of Y is correlated with the acidity of the extractant, while Du Preez and Preston attributed the changing order to steric hindrance caused by the structure of the carboxylic acids and the atomic number of rare earth metal ions. With straight chain and non-hindered acids the behavior of yttrium most closely resembles that of light lanthanides (e. g., Ce, or Pr), while for the sterically hindered acids the behavior of yttrium most closely resembles that of middle lanthanides (e. g., Gd or Tb).

Naphthenic acid has been widely used for separating yttrium from lanthanides in China. However, the extractant composition changes after being used for some time, and its high solubility in water leads to significant reagent losses. Novel carboxylic acids, such as sec-nonylphenoxy acetic acid (CA-100) and sec-octylphenoxy acetic acid (CA-12), developed in China, have much lower aqueous solubilities than naphthenic acids. A study on the extraction of trivalent lanthanides (Sc, Y, Ln) and divalent transition metal ions (Cu, Zn, Ni, Mn, Cd, Co) from acidic chloride solutions with CA-100 in heptane indicated that CA-100 can extract rare earth ions at lower pH values than Versatic 10. The extraction behavior of yttrium with CA - 100 most closely resembled that of the heavy lanthanides. The extraction behavior of trivalent rare earths using cekanoic, neo-heptanoic, and some 2-bromo alkanoic acids was also investigated. It was suggested that the introduction of the 2-bromo substituent in an alkanoic acid structure lowers the pKa values, enabling the substituent acids to be effective extractants at lower pH than the parent compounds.

2. 1. (1) 2 Organophosphorous acids

Various kinds of acidic organophosphorous extractants have been used in rare-earth separation processes, (*of which*) with D2EHPA (or HDEHP, di (2 - ethylhexyl) phosphoric acid) and HEHEHP (or EHEHPA, 2-ethylhexyl phosphonic acid mono-2-ethylhexyl) (*are*) being most widely used. (*The pioneering work of*) Peppard et al. (*used*) noted that the distribution coefficients of rare earth ions at tracer concentrations between D2EHPA in (toluene and (*s the diluent) to extract rare earth metals from their*) aqueous chloride solutions had an inverse third - power dependency on the HCl concentration in the aqueous phase(*. Their result indicated that at low metal loadings, the distribution coefficients of rare earth ions with D2EHPA had direct*) and a third-power dependency on the D2EHPA concentration in the organic phase (*and an inverse third-power dependency upon the HCl concentration in the aqueous phase*). Thi(*u*) s indicates that only one of the acid groups in a D2EHPA dimer in the organic phase dissociates and participates in the extraction reaction ((*as shown in*) following e(*E*) quation (1) 4). (*At* ) Gels formed in the organic phase at high metal loadings and(*, gel formation in the organic phase was observed at*) low acidities, which is(*were highly*) undesirable because of the associated viscosity and phase separation problems.

(*When*) The selectivity order for extracting rare earths from 0. 5M HCl solution with 0. 75M D2EHPA in toluene was(*, a selectivity order of*) Lu>Yb>Tm>Tb>Eu>Pm>Pr>Ce>La (*was observed*) (Figure 2), with the log of the distribution coefficient (*A plot of Log D*) ((*defined as*) called l(*L*) og K (*in their work*) by Peppard et al.) increasing linearly with(*as the function of*) the atomic number, Z, of the rare earth. The average (*was represented by a straight line of positive slope corresponding to an average value of 2. 5 for the*) separation factor of two adjacent rare earth elements was 2. 5. (*In same*) Yttrium was extracted between Tb and Tm in this solvent extraction system, (*the extraction of yttrium falls between that of Tb and Tm* corresponding(*lose*) to (*that of*) an artificial atomic number 67. 6. The extraction of (*all*) the lanthanide elements with D2EHPA in toluene was similar for(*from the*) perchloric acid solutions (*by the solvent mixture of D2EHPA in toluene exhibited similar results (the selectivity order was: La>Ce>Pr>Nd>Pm>Sm>Eu>Gd>Tb>Dy>Ho>Y>Er>Tm>Yb>Lu)*), 【批注】: I deleted this sequence because it appears to be the wrong way round. Higher Z gives higher extraction but poorer in nitrate media. It should be pointed out that although (*generally*) the distribution coefficients of rare earths generally increase(*s*) with increasing atomic number (*in the rare earth series*), the precise separation factors (*vary to different extents* with) depend on the acidity of the (*and*) aqueous (*media*) phase and nature of the anion.

In 1965, Molycorp demonstrated t (*T*) he large scale application of D2EHPA for pre－concentrating europium to about 15% from rare earth chloride feed derived from bastnesite, (*where it*) containing(*ed*) about 0. 1% $Eu_2O_3$(*was demonstrated by Molycorp in 1965*). Preston and du Preez pre–concentrated europium from chloride containing 0. 22 ±0. 01M total rare earths (%: Eu 93, Sm 3, Nd 2, Ce 1, Pr 0. 5 and Gd 0. 5) using 0. 4M D2EHPA in xylene; 99. 98% Eu solutions were obtained in a single extraction stage at pH 2. 7. 1M D2EHPA was used to separate a 99. 8% $La_2O_3$ product from didymium chloride solution (feed containing 45% $La_2O_3$, 35% $Nd_2O_3$, 10% $Pr_6O_{11}$ and 5% $Sm_2O_3$). (*The process for separation and purification of lanthanum from didymium chloride solution using D2EHPA was also described. It was reported that a product of about 99. 8% $La_2O_3$ could be obtained from the feed containing 45% $La_2O_3$, 35% $Nd_2O_3$, 10% $Pr_6O_{11}$ and 5% $Sm_2O_3$ by using 1mol/L D2EHPA in AmSCO mineral spirit (from American Spirits Company) through two*) Two multistage counter–current extraction circuits w(*h*) ere needed; (*in*) the first, 12 stage cascade ((*12 stages*), *the La was*) concentrated La in the raffinate, and (*in*) the second, 14 stage cascade ((*14 stages*), *the La was*) provided further purification(*ed*). The over–all recovery of La was 60%. (*Preston and du Preez (1996) used D2EHPA (0. 4mol/L in xylene) for pre–concentrating europium from a solution of the chloride containing 0. 22 ±0. 01mol/L total rare earths (%: Eu 93, Sm 3, Nd 2, Ce 1, Pr 0. 5 and Gd 0. 5). They reported that under nitrogen protection and pH 2. 7 with an equal volume of two phases, upgraded europium solutions with purities in excess of 99. 98% were readily obtained in a single extraction stage.*)

It was reported that the separation of rare earths with D2EHPA in nitrate media was poorer than in chloride. (*A*) Preston et al. described a continuous solvent extraction process for separating the middle (Sm, Eu, Gd, and Tb) and the light rare earth fractions (La, Ce, Pr, and Nd) from a nitrate feed (*liquor was described by Preston, et al*). The (*process consisted of extracting the*) middle rare earths were first extracted into a 15% v/v of D2EHPA in Shellsol AB in an 8 (*eight*) – stage (*of*) counter–current circuit, followed by scrubbing with 1mol/L $HNO_3$ in (*two*) 2–4 (*to four*) stages, and stripping with 1. 5mol/L HCl in (*six to eight*) 6–8 stages. Residual rare earths in the organic phase (*which were*) mainly Dy, with some Tb and Gd) were removed in a secondary stripping circuit using 2. 5mol/L HCl in four stages. (*More than*) Over 1000L of feed liquor (*were*) was processed in two continuous counter–current trials lasting a total of 630 hours. From (*the*) feed containing Sm: 3. 5g/L, Gd: 2. 4g/L, Eu: 0. 8g/L, and Nd: 20g/L (together with 4 to 8g/L each of the lighter rare earths), strip liquors containing Sm: 35g/L, Gd: 20g/L, and Eu: 8g/L were obtained with neodymium (5g/L) as the main impurity. (*Recoveries*) The recoveries of the middle rare earths to the strip liquors were relatively high (95%–100%), whereas losses of the light rare earths were (*relatively*) low (<4%). D2EHPA has also been used to separate Sm, Eu, and Gd from the other rare earths in a mixed nitrate–chloride leachate from monazite (*The process based on extraction by the solvent mixture of D2EHPA in kerosene from nitric–hydrochloric acid mixture was developed*). (*The monazite concentrate (93. 11% $RE_2O_3$) was digested with mixture of HCl and $HNO_3$ (the volume ratio of 1:1) after removal of most of Ce. The solvent mixture of D2EHPA in kerosene was used* *to separate Sm, Eu, and Gd from the other rare earths in the leachate. A final middle rare earth concentrate containing 60.6%, 3.21% and 34.4% of Sm, Eu, and Gd, respectively, with 1.66% Nd and 0.02% Tb in their oxides was obtained with about 78% recovery.)*

It is known that D2EHPA can extract rare earth at (*significantly*) low pH values, but because the equilibrium of equations 1 and 2 lies strongly to the right it is (*however, it has the disadvantage of*) difficult to strip the loaded metals. Thus, other acidic organophosphorus extractants have been widely examined for rare earth solvent extraction. Benedetto et al. reported that DS5834 ((*developed by*) Zeneca Specialties, with (*and has*) a formulation (*of a type*) similar to M2EHPA, mono-2-ethylhexyl(*e*) phosphoric acid) could effectively extract Ga, In, Sm, (*gallium, indium, and samarium*) and Gd(*gadolinium*) from acidic media but was (*not*) neither selective for Sm and Gd, nor(*the rare earth elements samarium and gadolinium, and consequently was not*) effective for the separating(*on of*) these metals. The reagent HEHEHP, marketed variously as PC-88A, SME 418, Ionquest 801 and P-507, has gained more popularity for rare earth separations (*since the loaded*) because rare earths (*on it*) can be (*more readily*) stripped at lower acidities than from D2EHPA. In addition, HEHEHP (*allows the extraction to be operated under a higher concentration of rare earths, because it*) can be more heavily loaded with rare earth than D2EHPA before the onset of saturation effects, which . (*This results in an*) increases (*of*) the extraction efficiency. The extraction of rare earths with HEHEHP (*was reported to*) follows equation 1, with the extraction (*the usual cation exchange mechanism and same extraction efficiency*) order (*for extracting rare elements*) from chloride and nitrate media of(, ) La < Ce < Pr < Nd < Sm <Eu < Gd < Tb < Dy ( ~Y) < Ho < Er < Tm < Yb < Lu(, *which increases with increasing atomic numbers, was observed for both D2EHPA and HEHEHP. A similar tendency was also observed in the extraction from nitric acid solutions.*) Balint investigated separation factors between adjacent rare earths extracted from a mixed rare earth chloride solution using HEHEHP and D2EHPA, (*respectively,* ) and reported that HEHEHP (*the former gives better separation*) is superior for samarium and heavy rare earths (*than the later*). (*The*) A process developed by Daihachi for separating rare earths using HEHEHP (*and the reagent*) has been applied in commercial separation plants in Baotou, China. Fontana and Pietrilli ((2009)) also suggested the use of HEHEHP for recovering rare earths resulted from (*the*) spent NiMnH (*spent*) batteries.

Some dialkyl phosphinic acids have also been investigated (*were also used*) for rare earth separation, (*of which*) although only Cyanex 272 (bis(2, 4, 4-trimethylpentyl) phosphinic acid) has (*found*) been used commercially (*applications*). Sabbot and Rollat described the(*A process for*) preparation of pure $Yb_2O_3$ (99.3%) from a mixture of ytterbium and lutetium oxides ($Yb_2O_3$ = 87.5%, $Lu_2O_3$ = 12.5%) using 1mol/L Cyanex 272 in kerosene (*from a mixture of ytterbium and lutetium oxides ( $Yb_2O_3$: 87.5%, $Lu_2O_3$: 12.5%) was described by Sabbot and Rollat*). Saleh et al. investigated the extraction (*behavior*) of La(Ⅲ) by Cyanex 272 in toluene from acidic nitrate-acetato media and(*um by Cyanex 272 in toluene. They*) suggested the formation of $La(Ac)_2A \cdot 3HA$ (Ac denotes acetate ion and HA denotes the acidic form of Cyanex 272) in the organic phase at (*a*) low(*er level of*) La loading and $LaA_3$ at higher loading. (*However, s*) Studies on the extraction of samarium from chloride solutions with Cyanex 272 indicated that the extracted species was $Sm(OH)A_2 \cdot 2HA$. *(The)* Studies on the extraction *(behavior)* of *(lanthanum(Ⅲ), praseodymium(Ⅲ), europium(Ⅲ), holmium(Ⅲ), and ytterbium(Ⅲ))* La, Pr, Eu, Ho and Yb into chloroform solutions containing dicyclohexylphosphinic acid (DCHPA) *(was also investigated. It was found)* showed that the extraction selectivity of DCHPA was inferior to that *(exhibited by)* of other dialkyl phosphinic acids, presumably *(which probably)* because the cyclohexyl *(sub-stituents)* groups in DCHPA *(lead to)* sterically hinder*(rance in)* chelate formation*(, thus decreasing the ligand selectivity)*.

(出于对文章结构的考虑，原稿 2.1.1 Carboxylic acids 原在此处，修改后将篇幅移至前面)

## 2.2 Chelating extractant

Being as hydrogen ion donors, chelating extractants *(can)* extract metals by a cation ion exchange mechanism *(according to the following reaction)* similar to equation 1, but but the resulting organic complexes are stabilized by the organic anion coordinating the central cation in at least two positions: (公式略)

*(where $M^{n+}$ denotes a cation of metal and HA denotes an acidic monobasic extractant in free form.)*

Chelating extractants have been examined for extracting europium from nitrate *(No chelating extractants proved suitable for rare earth separations though some has been tested on extraction of rare earths from aqueous)* solutions, and . *(The extraction of)* cerium(Ⅲ) and lanthanum(Ⅲ) from chloride solutions *(with LIX 54 (Cognis; the active component is dodecylphenyl-methyl-β-diketone) was examined. Though cerium was better extracted by the extractant than lanthanum, the selectivity of their separation was unattractive)*, but performed unfavorably compared *(to those)* with acidic extractants.

## 2.3 Solvation extractant

Several types of solvation extractants have been used for rare earth separations. *(In the early work of)* Peppard et al. investigated the extraction of trivalent rare earths from chloride and nitrate solutions with TBP (tributylphosphate) *(was investigated)*. The extractability of the lanthanides with TBP increased with increasing atomic number, but the distribution coefficients *(in the chloride systems)* were much lower in chloride solutions than *(those)* in nitric *(systems)* media. *(The)* Concentrated nitric systems were *(more)* promising for separating rare earths lighter than samarium*(, provided the nitrate activity was high)*. Rare earths heavier than samarium *(can )* could not be separated effectively *(separated)* in nitric systems. The rare earths in neutral nitrato complexes are coordinated by the phosphoryl group of TBP, yielding an extractable complex. The overall reaction can be expressed as: (公式略)

*(although)* There would be few simple $Ln^+$ cations in solution at the ionic strengths needed for effective extraction *(($\overline{TBP}$ and $\overline{Ln(NO_3)_3(TBP)_3}$ denote TBP and the complex formed by the lanthanide with TBP in the organic phase, respectively). The)* Later work of Peppard, et al. examined

the influence of extraction conditions on the equilibrium constant of reaction 5 to infer the composition of the complexes formed under different conditions (*discussed details on the mechanism of extraction of rare earth in nitric media. Plot of Log K (K: the extraction equilibrium constant of reaction 6) vs. Z (Z: atomic number) approximates two straight lines joining at Z = 64. The line for the high-Z region has a positive slope somewhat less than that of the line for the low-Z region at 18.5 M $HNO_3$. This difference in slope becomes intensified as the concentration of nitric acid is lowered, the high-Z line ultimately acquiring a negative slope. The slope of the low-Z line remains positive throughout the region studied. They suggested that the extracting species is $Ln(TBP)_a(H_2O)_{x-a}(NO_3)_3$, in which in which x is probably 6, but is smaller in the high-Z than in the low-Z range and a is a function of both the nitrate activity and atomic number Z (e.g., in the neighborhood of 2mol/L $HNO_3$, 2 for all Z values and, in the neighborhood of 16mol/L $HNO_3$, 3 for low-Z nuclides and 4 for high-Z nuclides).*) Lu et al., studying the s(*S*)olvent extraction of Ce(Ⅳ) and Th (Ⅳ) from sulfate solutions with Cyanex 923 in n-hexane, found the extraction of Ce(Ⅳ) to be insensitive to acidity, while the extraction of Th(Ⅳ) increased with the aqueous acidity (*from sulfuric acid medium was studied by Lu et al. They reported that the aqueous acidity exhibit little effect on the extraction of Ce(Ⅳ) but the extraction of Th(Ⅳ) increases with the aqueous acidity.*) A third phase formed at (*moderate to strong sulfuric acid (e.g.,*) $H_2SO_4$>5mol/L(), *a third phase was formed*). The extraction (*mechanism*) of Ce(Ⅳ) and Th(Ⅳ) from sulfate media with Cyanex 923 can be represented by the reaction: (公式略)

(*where*) M represents Ce or Th and B denotes Cyanex 923. The extraction of trivalent lanthanides and yttrium from nitrate medium with Cyanes 925 in heptane was (*studied by Li, et al*) suggested to follow the reaction. (*The extraction mechanism was suggested as follows*): (公式略)

where M and B represent the metal and Cyanex 925, respectively.

During the 1960s, Thorium Limited in the United Kingdom used TBP to separate light rare earths in nitrate media (*during the 1960s*). This process was operated batchwise with total reflux; on attaining steady state, the process was stopped and products of different composition were withdrawn from different stages. (*The process appears to be*) This configuration is costly compared to (*the*) continuous processing and c(*an*) not amenable to(*be operated in a large*) scale up. Preston et al. described (*the development and implementation on*) a pilot-scale (*of a*) process for (*the*) recovering(*y of*) a mixed rare-earth oxide product from (*the*) calcium sulfate hemihydrate sludge (*that aroused*) generated during the manufacture of phosphoric acid from apatite mined at Phalaborwa(*, in*) South Africa. (*The r*) Rare earths were recovered from (*the*) leach liquor containing 1.0M nitric acid and 0.5M calcium nitrate by (*the*) additi(*o*) ng (*of*) 2.5M ammonium nitrate and extracting (*on*) into 33% v/v DBBP (dibutyl butylphosphonate) in Shellsol 2325. The organic phase was stripped with water to yield a solution of rare earth nitrates from which the mixed rare earth oxide was recovered by (*the*) adding(*tion of*) oxalic acid and calcining(*ation of*) the precipitate. Later work examined using TBP (15% in Shellsol K diluent) to selectively extract cerium(Ⅳ) from the rare earth nitrate feed (*In a continuous counter-current laboratory trial of the process, a total of 140kg of sludge was processed to produce 265 L of leach liquor, which was treated in five extraction* *and five stripping stages to give strip liquor from which 4 kg of mixed rare earth oxide of 98% purity was recovered. In their later work.* ) The organic phase was stripped by reducing the cerium( Ⅳ) with dilute hydrogen peroxide in two stages, giving solutions containing up to 90g/L of cerium( Ⅲ) . (*TBP (15% in the Shellsol K) was used to selectively extract cerium( Ⅳ) from the rare earth nitrate feed. In continuous counter-current trials, four extraction stages were used, followed by four stages of scrubbing with 3mol/L nitric acid. Stripping of the organic phase was accomplished by reduction of the cerium( Ⅳ) with dilute hydrogen peroxide in two stages, giving solutions containing up to 90g/L of cerium( Ⅲ). Addition of oxalic acid to the strip liquors, followed by calcination of the precipitated oxalate, gave cerium dioxide of at least 99.98% purity in about 70% yield.* )

## 2.4 Anion exchangers

Anion exchangers extract metal ions as anionic complexes, and hence are only effective in the presence of strong anionic ligands. Early (*Some earlier* ) work indicated that the separation factors for adjacent rare earths (*are not especially attractive when*) with primary (*amines*) or tertiary amines (*was used for extracting rare earths from*) are poor in chloride media. Sulfate media are more promising. El-Yamani and Shabana (*studied*) suggested that the extraction (*behavior*) of lanthanum ( Ⅲ) from (*sulfuric*) —sulfate solutions with Primene JMT ( tri-alkyl methylamine) (*in kerosene. They suggested the extraction mechanism as*) follows the reactions: (公式略)

where $\overline{RNH_2}$ denotes the Primene JMT in the organic phase. (*The investigation on the liquid-liquid extraction of*) Y( Ⅲ) (*from sulfate medium using Primene JMT further confirmed this extraction mechanism*) behaves similarly. Quaternary ammonium salts such as tri-octyl methylammonium nitrate (Aliquat 336) have proved promising for separating(*on of*) rare earths. (*Quaternary ammonium compounds with high molecular such as Aliquat 336 behave chemically as strong-base anion exchangers and require lower concentrations of salting out reagents. Extraction and separation of rare earth pair was improved by addition of EDTA as the chelater.* ) The (*mechanism of* ) extraction (*of rare earths with quaternary amines*) reaction can be (*expressed* ) represented as (*following,* ): (公式略)

where Ln denotes the rare earth ion and $\overline{R_4N^+NO_3^-}$ the quaternary ammonium nitrate salt, although Cerna et al. suggested a. (*A*) more complicated reaction (*mechanism was suggested by Cerna et al.* ). These reagents are strong-base anion exchangers and require lower concentrations of salting out reagents than amines. Chelation with EDTA improved the extraction and separation of rare earth pairs. In nitrate media Aliquat 336 extracts light rare earths more readily than the heavier ones. This behavior contrasts that of most of cation exchange and solvating extractants, for which the extraction of the rare earth metals increases steadily with increasing atomic number. Hence, quaternary ammonium salts provide a means of removing light rare earths from process solutions. (*The selectivity of individual rare earth by Aliquat 336 is low and the heavy rare earths are extracted more strongly in thiocyanate media than in chloride media. In nitrate media Aliquat 336 extract light rare earths more readily than the heavier ones. This behavior contrasts sharply with that of most of extractants commonly used in rare earth separations, such as organophosphorus and carboxylic acids, for which the extraction of the rare earth metals increases markedly with increasing atomic number. Extraction by quaternary ammonium salts thus provides a means by which the rare earths of lower atomic number can be removed from the predominantly neodymium-containing light rare earth fraction remaining after the extraction of the middle and heavy rare earths with an organophosphorus acid.)*

Yttrium is anomalous, behaving as a heavy rare earth in nitrate media and as a light in thiocyanate media. This has been utilized for treating xenotime, which is about 60% $Y_2O_3$ (Table 2). Xenotime concentrate was leached with $HNO_3$ and lighter rare earths were extracted with Aliquat 336 in an aromatic diluent. Yttrium and the heavy rare earths remained in the aqueous phase. Yttrium was then extracted by Aliquat 336 from a thiocyanate solution, yielding 99.99% $Y_2O_3$, while other heavy rare earths remained in the raffinate.

(*In an early patent, Guadernack described a solvent extraction process using Aliquat 336 for separating yttrium from lanthanides. The raw material used was 60% $Y_2O_3$ concentrate extracted from xenotime. The concentrate was first dissolved in $HNO_3$ and lighter rare earths were extracted by using quaternary ammonium compounds dissolved in aromatic diluents. Yttrium and heavy rare earths remained in the aqueous phase and were subject to a second solvent extraction process through which yttrium was extracted by the quaternary amine extractant from thiocyanate media while other heavy rare earths remained in the raffinate. Yttrium was upgraded to 99.99% $Y_2O_3$ by these solvent extraction procedures.*) Lu, et al. isolated >99% $Nd_2O_3$ (*used 45-stage tube-type mixer-settler for extraction of neodymium*) from didymium nitrate solution ((*a mixture of*) 83% Nd, 15% Pr and 2% other rare earths) (*by quaternary amine* (*Aliquat 336*). *$Nd_2O_3$ with a purity >99% could be obtained and 95% of*) with 95% Nd (*was*) recovery(*ed*) using Aliquat 336 in a 45-stage extraction. Preston described (*report*) ed a solvent-extraction process for recovering neodymium oxide (95% $Nd_2O_3$) from (*the*) light rare earth nitrate solution using 0.50M solution of Aliquat 336 nitrate in Shellsol AB. (*In a continuous counter-current mini-plant trial, 190 L of feed liquor containing 16-17g/L Nd, 3.5g/L Pr, 7g/L Ce, and 7.5g/L La were processed in two separate passes with a combined duration of 220 h. In the first pass, 95% of the lanthanum and 75% of the cerium contained in the feed liquor were removed, whilst in the second pass the residual cerium and most of the praseodymium were extracted, leaving a purified neodymium solution as the raffinate. Both circuits used a 0.50 M solution of Aliquat 336 nitrate in Shellsol AB in 8 extraction and 6 scrubbing stages. Stripping of the loaded organic phase was carried out with water in 6 stages, part of the resulting strip liquor being recycled as the aqueous feed to the scrubbing circuit. Under conditions of sufficient reflux of strip liquor, the raffinate from the second pass contained up to 75% of the neodymium present in the original feed at a purity of 95-96% $Nd_2O_3$ (with 2% $Pr_6O_{11}$; 1.5% $Sm_2O_3$; 0.7% $CeO_2$; and 0.2% $La_2O_3$). From the strip liquors produced in the first pass, a lanthanum concentrate containing 51% $La_2O_3$ (with 36% $CeO_2$; 7% $Pr_6O_{11}$; and 6% $Nd_2O_3$) was obtained, whilst from the strip liquors produced in the second pass a praseodymium concentrate containing 32% $Pr_6O_{11}$ (with 41% $Nd_2O_3$; 22% $CeO_2$; and 4% $La_2O_3$) was obtained.*)

## 2.5 Synergistic solvent extraction

Numerous types of synergistic solvent extraction systems for extracting and separating rare earths have been reported, (*of which*) including mixtures of acidic extractants (e.g., carboxylic or organophosphorus acids), mixture of neutral extractants (e.g. TBP and TOPO), and combinations of these (*two types of extractants have been used*). Preston and du Preez examined the effect of the addition of a series of tri-alkyl phosphates $(RO)_3PO$, di-alkyl alkylphosphonates $(RO)_2RPO$, alkyl di-alkylphosphinates $(RO)R_2PO$, and tri-alkylphosphine oxides $R_3PO$ on the extraction of the trivalent lanthanides and yttrium from chloride media by DIPSA (3,5-di-isopropylsalicylic acid). (*To different extents, s*) Synergistic effects were (*was*) observed with all (*the*) mixtures, albeit to different extents. For the series of compounds with $R=n$-butyl, the synergistic effect increased in the order $(RO)_3PS < (RO)_3PO < (RO)_2RPO < (RO)R_2PO < R_3PO$. The synergistic effects were greater for lutetium(Ⅲ) than for lanthanum(Ⅲ) (the separation across the lanthanide series increased). Mixtures of DIPSA and TIBPO (tri-isobutyl phosphine oxide) gave somewhat better separation factors between the light and the middle lanthanides ($\beta_{Sm/Nd}=3.0$) than Versatic 10 acid alone ($\beta_{Sm/Nd}=2.6$). Separation factors were comparable to those with the latter extractant between the heavy lanthanides (thulium to lutetium). (*They* ) The authors suggested that the extracted rare earth complexes had a composition of $LnA_3L_2$ (where HA represents carboxylic acid and L the neutral organophosphorous compound) and the synergism(*tic effect was*) resulted from the replacement of some or all the undissociated carboxylic acid molecular (see equation (2)4).

The comparison of the effects of some bi-functional ligands containing C=O, P=O or S=O groups upon the extraction of trivalent rare earth metals from chloride media by DIPSA in xylene indicated that the shifts generally increased in the order S=O<C=O<P=O for comparable (*bi-functional compounds containing these donor groups*) ligands. The synergistic effect produced by the addition of a given bi-functional compound generally decreased across the lanthanide series (La to Lu), attributed (*which was believed*) to (*be due the*) steric hindrance effect. Reddy(, ) et al. (*investigated the synergistic extraction of rare earths from nitrate solutions with mixture of Cyanex 301 and Cyanex 923. They*) reported that (*in the presence of Cyanex 923,* ) La(Ⅲ) and Nd(Ⅲ) were extracted from nitrate solution by Cyanex 301 (HA) and Cyanex 923 (L) as $LnA_2 \cdot NO_3 \cdot L$ (*where HA = Cyanex 301 and L = Cyanex 923*), while Eu(Ⅲ), Y(Ⅲ) and heavier rare earths were extracted as $LnA_3 \cdot HA_2 \cdot L$. T(*he addition of the t*) ri-alkylphosphine oxide (*to the extraction system not only*) significantly enhanced both the extraction efficiencies (*of these metal ions but also improved the*) and selectivitiesy (*significantly*), especially between yttrium and heavier lanthanides. Zhang et al. reported studies of t(*T*) he solvent extraction of (*Cerium*) cerium(Ⅳ) and (*Fluorine*) fluoride from sulfate(*uric*) solutions using a mixture of Cyanex 923 and D2EHPA in n-heptane (*have been investigated by Zhang, et al.. They reported that Ce(Ⅳ), $CeF_2^{2+}$ were effectively extracted by the solvent mixture and separated from complexes as $Ce(HSO_4)_2A_2 \cdot 1.5B$ and $CeF_2(HSO_4)A_2B$ (where HA = D2EHPA and B = Cyanex 923), respectively. However,* ) Ce(Ⅲ) was not extracted by the mixture. (*High separation factors between Ce(Ⅳ), F and Ce(Ⅲ) (Ce(Ⅳ)/Ce*

(Ⅲ) = 1321.18, $CeF_2^{2+}/Ce$(Ⅲ) = 36.35) *were obtained.*)

(*For the*) Binary acid extractant systems, Ying(, ) et al. examined the extraction of $Yb^+$ from chloride solution with Cyanex 272, P507 (HEHEHP), and their mixtures in n-heptane. The extraction of(*y reported that*) $Yb^+$ was (*extracted as* $YbA_2L_4H_3$ (*HA* = *Cyanex 272; HL* = *P507*) *in the organic phase. The extraction ability of* $Yb^{3+}$) higher with the mixture (*was higher*) than (*that*) with Cyanex272 or P507 alone. (*Synergistic*) A synergistic effect was observed on the separation of Yb/Tm and Lu/Yb, but not (*of*) for Tm/Er, Er/Ho, and Ho/Dy. Zhang, et al. investigated the extraction of (*RE*(Ⅲ) (*RE* =) trivalent La, Nd, Sm, and Gd) from sulfate(*uric acid*) media by (*the*) a mixture of D2EHPA and HEHEHP. A s(*S*) ynergistic effect was observed for the extraction of all four metals at pH = 2.0. (*They suggested the extraction of rare earth ions with the mixed extractants as following:*) (公式略)

(*where HA and HL represents D2EHPA and HEHEHP, respectively. T*) Li et al. also reported that the separation factor for Sm and Nd was significantly increased using a mixture of D2EHPA (40% v/v) and HEHEHP (60% v/v) (*extraction of Nd and Sm from sulfuric acid media with the mixture of D2EHPA and HEHEHP has been examined by Li, et al.. The separation factor of Sm and Nd was significantly increased when the mixture of D2EHPA (40% v/v) and HEHEHP (60% v/v) was used.*)

Sun, et al. examined the synergistic extraction of trivalent (*rare earths* (*RE* =) Sc, Y, La, Gd, and Yb) from (*hydro*) chloride media(*um*) using a mixture of Cyanex 272 and sec-nonylphenoxy acetic acid (CA-100) in n-heptane. T (*and suggested the extraction reaction of yttrium with the mixed extractants as following:*) (公式略)

(*Where HA and HL represents CA-100 and Cyanex 272, respectively. They found that t*) he separation factor for(*raction of*) Yb and Y was much higher than that (*of the*) with CA-100 alone. The extraction of rare earth elements from chloride (*medium*) media by mixtures of CA100 (*and*) with Cyanex 301 (*and mixture of CA100 and*) or Cyanex 302 (*has*) was (*been*) studied by Tong(, ) et al. In the CA100+Cyanex 301 system, the synergistic enhancement coefficients decreased with increasing atomic number of lanthanoids, but t(. *However, such an order could not be found in the CA100+Cyanex 302 system. T*) he separation factors between Y and all the lanthanoids were enhanced. (*by the CA100+Cyanex 301 system.*) Jia, et al. reported that the separation factors of all adjacent trivalent rare earths were better in a mixture of sec-octylphenoxy acetic acid (CA12) and Cyanex301 in n-heptane than in Cyanex 301 alone (*examined the synergistic solvent extraction of RE*(Ⅲ) *with the mixture of sec-octylphenoxy acetic acid* (*CA12*) *and Cyanex 301 in n-heptane. They reported that separation factors of all adjacent rare earth elements with the mixture were improved markedly compared to those with Cyanex 301 alone*).

## 3 Process Engineering and Equipment

### (3.1 *Configurations*)

Although there is an extensive literature on rare earth solvent extraction chemistry and equilibria, as discussed above, far less is known on the engineering details of rare earth separations. From the limited open literature, rare earth producers appear to follow similar approaches. There is often a need for a primary separation of rare earths from impurities in the original leach solution, along with concentration. D2EHPA has been widely used for primary separation because the distribution coefficients of the rare earth elements as a group differ markedly from those of typical impurities in leach liquors. D2EHPA is also suitable for concentrating the rare earth elements from dilute, acidic solutions.

In general, rare earths are separated in the trivalent state. They are usually first separated into two, three or sometimes four groups, followed by subsequent separation of individual rare earth. Preferential separation of yttrium is desirable, and cerium and europium are often separated initially on the basis of different valance states ( $Ce^{+}$ and $Eu^{+}$).

The extractants and aqueous anion are generally selected considering both cost and technical requirements, and the impact on the process configuration . For example, cationic exchangers usually offer higher selectivity on rare earth ions compared to neutral and anionic exchangers. However, the reactive chemical requirement is greater with cation exchangers, because base is required to drive extraction, and acid is required for selectively washing the organic phase. In contrast, with solvation extractants and anion exchangers the reactive chemical requirement is lower. Thus there is a trade-off between selectivity (which lowers the number of stages, and hence capital and chemical inventory costs) and the operating costs.

*(Many papers and reports on the rare earth solvent extraction process have been published, however, the details of processes and operations in practice are usually unobtainable. According to the relevant information open to the public, rare earth producers usually follow almost identical principles or scheme on selecting process routes for solvent extraction separation of rare earths from a leachate or other aqueous solutions*[(Gupta.)]*. These overall principle or scheme includes (but not limited to):*

- *The separation of rare earth elements is generally performed in trivalent states;*
- *Rare earth elements are usually first separated into two, three or sometimes four groups followed by subsequent separation of individual rare earth;*
- *Preferential separation of yttrium is desirable;*
- *Selective extraction of cerium and europium using their valence states $Ce^{+}$ and $Eu^{+}$ is applicable in necessary;*

*In addition, the choice of extractants and aqueous phase is influenced by both cost considerations and requirements of the technical performance. For example, cationic exchangers usually posse higher selectivity on rare earth ions compared to neutral and anionic exchangers. However, when the rare earths are extracted using cationic complexes, the reactive chemicals requirement is more because a base is required for extraction from the aqueous phase and an acid is required for selective washing less extracted rare earths from the organic phase. Comparatively, in the neutral extraction media or when the rare earths are extracted as an anionic complex, the reactive chemicals requirement is less. From the viewpoint of this consideration, neutral or anionic extraction is more cost-effective. However, cationic extraction will be preferred when high selectivity is of major importance to the process.* )

## 3.1 Configurations

According to the objective, rare earth solvent extraction processes (*can be roughly identified as two classes*), are generally classified as primary separations, which aim(*s*) to separate rare earth elements from other elements (which is relatively straightforward, and comparable with other solvent extraction processes in hydrometallurgical operations), and secondary separations, which (*aims to separate*) produce single or mixed (typically 2 or 3) rare earth products from the mixed rare earth (*product obtained*) stream produced by (*from*) primary separations. The latter is often much more challenging, particularly when producing a single, pure rare earth, because of the chemical similarity of the rare earths. However (*It should be noted that sometimes*) a (*one*) single-step solvent extraction (*is*) may be enough to produce designated products. As mentioned (*previously*) above, D2EHPA has been widely used (*as the extractant for the*) in primary separation processes (*since*) because the distribution coefficients for the rare earth elements (*as a group* ) with D2EHPA are(*is*) markedly different from (*that*) those of (*the*) other elements in the aqueous solution (leach liquor). (*This extractant*) D2EHPA is also suitable for concentrating the rare earth elements from dilute solutions because of (*its high extraction ability for the rare*) earth element(*s, which enables it to extract them from dilute and often acidic solutions*) the high distribution coefficients.

(*On the aspect of equipment, up to*) A plant producing multiple single rare earth products may contain hundreds of stages of mixers and settlers (*may be assembled together to provide mixing and settling*) altogether. (*Three typical configurations for rare earth solvent extraction circuits have been summarized by Doyle, et al.*.) A (*common* ) classic countercurrent flow scheme for a simple solvent extraction circuit is shown in Figure 3. The aqueous stream leaving the $n$th mixer settler would be pumped to the ($n-1$) th mixer settler, w. (*W*) hile the organic phase (which is ideally in equilibrium with the aqueous phase leaving the $n$th stage) would be pumped to the ($n+1$) th mixer settler. This arrangement may also be used for simple rare earth separations, e. g., the primary separation of rare earths from impurities. However, there is fundamental shortcoming with this configuration if one wishes to separate rare earths from a mixed feed. (*is usually use for simple separation of rare earths, e. g., the separation of light REs and heavy REs, due to their low separation factors with most commercial extractants.*) Consider an operation using a cation exchange extractant or a solvating extractant(*s*), for which the heavier rare earths have a stronger affinity for the organic extractant phase than have the lighter ones. (*Hence,* ) I if mixed aqueous feed were introduced into one end of a bank of mixer settlers, the organic phase leaving that end would be somewhat enriched in heavy rare earths, but there would still be an appreciable amount of light rare earths in this stream, because of the low separation factors. This shortcoming is addressed by introducing the mixed aqueous feed near the middle of the bank of mixer settlers, as shown in Figure 4. A different aqueous stream is admitted into the end (the ($n+m$) th stage) to allow an appropriate number of stages for the light rare earth to be scrubbed from the organic phase back into the aqueous phase. This scrub solution may exit the process midway, or may continue on with the aqueous feed. To minimize the dilution of rare earth concentrations caused by the introduction of the barren aqueous scrub and the organic phase, reflux is sometimes used; : some of the light rare earths from the aqueous raffinate are loaded back into the organic phase entering the first (*miser*) mixer settler, and some heavy rare earths are stripped from the organic product and added to the aqueous scrub (Figure 5). When operating under reflux, the mass transfer occurring at each stage becomes an exchange of different rare earths, according to their affinity for an extractant, as exemplified below for a liquid cation exchange extractant. (公式略)

Reflux minimizes dilution and also reduces the number of stages need to effect a given separation. Nevertheless, (30-60) many separation stages are typically needed to obtain pure product. Most banks of mixer settlers are set up to make a single separation between two adjacent rare earths, but some configurations have (*. In some cases, separation process with*) three or more (*outlets of*) product streams (*is applied*). (*Different solvent extraction processed for separating and purifying rare earths from various resources have been documented in details in the work of Gupta (2005)*) . Some typical applications and process flowsheets for solvent extraction separation of rare earths used in practice are summarized (*at*) below.

*Molycorp-bastnasite*

Figure 6 shows t(*T*)he schematic flowsheet for the Molycorp process, used to extract (*for making*) europium oxide from the leachate of Mountain Pass bastnasite (*is shown in Figure 6. The rare earth*) A chloride solution (100g/L REO) containing all the rare earth except Ce (*originally was obtained after the ore was*) is generated by calcination(*ed*) and leaching(*ed*) with HCl solution. (*In the Eu recovery circuit, t*) Two steps of solvent extraction with D2EHPA were applied. The chloride solution was first (*extracted*) contacted with 10% v/v D2EHPA in kerosene, and the extraction was performed in (*through a*) five stages of mixers and settlers under conditions that "split" the rare earths with Sm and all heavier rare earths reporting to the D2EHPA solution, and Nd and all lighter elements reporting to the raffinate (this is designated in Figure 6 by "(Nd/Sm)" in the first solvent extraction stage). There are two engineering reasons for splitting the rare earths in this way initially. The first is that it is relatively easy to separate Nd and Sm because although they are consecutive elements in any natural rare earth minerals, they are not consecutive elements on the periodic table; the intermediate rare earth element, promethium, does not occur in nature. Thus their separation factor is typically double that of any other consecutive rare earth pair. The second reason is that, referring back to table 3, it is apparent that Sm and the heavier rare earths account for only a very small proportion of the rare earths in bastnesite. Thus they can be removed using a small volume of D2EHPA solution, and the resulting rare earth mixture can be further processed using much smaller mixer settlers, leaving La, Pr, and Nd, the bulk of the rare earths in the concentrate, in the aqueous raffinate. These were precipitated with ammonium and sodium hydrogen sulfide, then further processed in much larger-scale equipment. More than 98% of europium in the solution was extracted. (*More than 98% of Eu in the solution was extracted and the raffinate containing La, Pr, and Nd were precipitated with ammonia and sodium hydrosulfide.*)

The loaded organic (containing 98& of the Eu from the leach liquor) was(*ere*) stripped with 4mol/L HCl. The iron in the strip solution was precipitated out at pH 3.5, and. (*After iron removal,*) the clarified Eu-bearing solution (*was subjected*) proceeded to a second solvent extraction circuit, also using. (*Again,*) 10% D2EHPA in kerosene and (*another*) five stages of mixers and settlers (*were applied*). Europium and other heavy rare earths were loaded in the organic phase, with the (*and*) light rare earths remaining(*s*) in the raffinate, which was returned(*verted*) to the first (*SX*) solvent extraction circuit. The (*loaded*) europium (and other rare earths) were(*as*) stripped from the loaded D2EHPA with 5mol/L HCl solution and the strip liquor (*then*) was passed through a column of zinc amalgam to reduce Eu(Ⅲ) to Eu(Ⅱ). Sulfuric acid was added to (*the divalent europium solution to*) precipitate europium sulfate, which was (*further processed by*) then calcined(*ations*) to produce pure $Eu_2O_3$ (99.99%). After europium recovery, the strip solution still contained Sm, Y, and other heavy rare earths. Gadolinium was extracted (*from the europium barren strip*) with D2EHPA in a 10-stage (*mixer-settlers*) extraction circuit followed by a 5-stage scrub(*bing*). The raffinate was neutralized with soda ash to precipitate Sm and the heavy rare earths.

*Rhô (o) ne-Poulenc-monazite*

It was reported that Rhô (*o*) ne-Poulenc (*could*) had the capability of producing(*e*) all the rare earth elements (*in*) at a purity of >99.999% almost entirely by solvent extraction. (*The*) A schematic flowsheet of the (*separation*) process is shown in Figure 7. (*The m*) Monazite concentrate was first digested with NaOH. (*After filtration, t*) The rare earths reported to the solid residue as hydroxides, which after filtration were dissolved in HCl (*and*) or $HNO_3$(*, respectively*). (*The clarified*) After clarification, the resulting solution(*s resulted from these processes*) proceeded to a series of solvent extraction circuits to produce individual rare earth oxides. (*The solvent extraction stream in the c*) Chloride media were used to prepare(*yielded*) a mixture of rare earth compounds, such as dehydrated rare earth chlorides, which were used to produce misch metal. (*The stream in the*) N (*n*) itrate media (*was*) were used to produce individual rare earth oxides, e.g., in the first separation circuit, lanthanum (99.9995% $La_2O_3$) remained in the aqueous phase while (*the*) a mixture of Ce, Pr, Nd, Sm, etc. was loaded into the organic phase. Similarly, $CeO_2$ (>99.5%) was separated from Pr, Nd, Sm, Eu, etc., after removing lanthanum. (*A variety of*) Various extractants, including carboxylic acids, organophosphorous acids, neutral organophosphorous compounds, and quaternary amines have been used in these separation processes. (*Rhone*) Rhône-Poulenc (*can*) could also produce high-purity individual rare earth oxides (*starting from not only monazite, but also*) from bastnesite or euxenite. The (*Rhone*) Rhône-Poulenc solvent extraction process has been regarded as the standard for all industrial producers.

*AS Megon-xenotime*

AS Megon developed a process for producing high-purity yttrium oxide starting from the xenotime concentrate. The schematic flowsheet is shown in Figure 8. The solvent extraction circuit consisted of a selective extraction by D2EHPA followed by three scrubbing and stripping units. The light rare earths ( La, Ce, Pr and partial Nd) and impurities including $Fe^{+}$ ions remained in the raffinate. The extracted yttrium and other rare earths were separated into three groups by selective washing. (*The obtained*) Y(*y*) ttrium nitrate solution was fed to the second circuit using the nitrate of tri-capry(*y*) l methylamine as the extractant. The lighter rare earths ( La, Ce, Pr, Nd and Gd, Tb, Er) were extracted while (*yttrium and*) Y, Tm, Yb, and Lu remained(*s*) in the raffinate, which was fed to the (*last*) third solvent extraction circuit, which used(. *A solvent extraction system of*) tri-capr(*y*) yl methylamine-$NH_4SCN$ (*was used*) to produce high-purity yttrium oxide.

*Mintek- apatite*

(*The recovery of r*) Rare earths have been recovered (*by solvent extraction*) from the calcium sulfate sludge (*resulted from the*) generated during the production of phosphoric from apatite at(*acid plant processing*) Phaleborwa (*apatite was reported*). Figure 9 shows t(*T*) he schematic flowsheet for the pilot plant (*using TBP as the extractant to produce mixed rare earths is shown in Figure 9*). The sludge was leached with dilute nitric acid solution (*with the addition of*) containing calcium nitrate. (*From*) Rare earths were extracted from the leachate(, *rare earths were extracted*) with TBP (40% v/v in Shellsol 2325). The raffinate was recycled back to leaching after (*the*) removing(*al of*) entrained organic solvent. The loaded organic (*solvent*) solution was stripped with water to yield a (*stream of*) mixed rare earth nitrate aqueous solution that was treated with ammonia and oxalic acid to precipitate(*from which*) a mixed rare earth oxalate (*precipitate was obtained by the addition of ammonia and oxalic acid*). This was calcined to give a(*e*) mixed rare earth oxide (89-94% purity) (*was produced by calination of the oxalate precipitate*). The rare earth oxide contained considerable amount of the middle rare earths, particularly Nd, Sm, Eu, and Ga. In subsequent pilot tests, TBP, HDEHP and Aliquat 336 (*have been*) were used to produce different rare earth products from the mixed oxides.

*Industrial processes in China*

The Shanghai Yue Long Chemical Plant, was reported to treat monazite concentrates in (*operate*) a (*similar*) process similar to the (*as*) Rhô (*o*) ne-Poulenc (*for treating monazite concentrates*) process. The simplified flowsheet of this plant is shown in Figure 10. After digesting the monazite in NaOH, filtration and leaching of the residue with HCl, t(*T*) he resulting rare earth chloride solution (*produced after HCl dissolution*) was extracted with D2EHPA and the rare earths(*elements*) were split(*divided*) into three groups, from which mixed and pure oxides, (*which were used to produce mixed rare earth oxides* ( ) carbonate, or chlorides were produced( ) *or individual rare earth oxides through extra solvent extraction circuits.* ) T(*For t*) he ion-adsorption type rare earth ores are first leached with(, ) HCl or $H_2SO_4$ (*solution was usually used to digest the ore first*). Cation exchange (*The acidic*) extractants, such as HEHHP and n(*N*) aphthen(*t*) ic acid, (*were commonly*) are frequently used (*applied* ) to extract rare earths elements from the (*resulted*) leachate, since these ores have high levels of heavy rare earths, (*elements in these ore usually occur as heavy rare earths*) which have a strong(*good*) affinity for(*with*) acidic extractants. Individual rare earth compounds (oxides or chlorides) can be produced through controlled(*ing*) stripping (*conditions*).

(*For the*) bastnesite ore, which is the main (*resource for*) rare earth resource in China, the ore or concentrate is usually (*process of*) roasted(*ing the ore or concentrate*) with $H_2SO_4$, (*are usually applied*) followed by leaching with water or dilute sulfuric acid. (*The r*) Rare earth elements (*in resulted solutions can be*) are recovered from the leachate by(*through*) solvent extraction with P204 (D2EHPA). (*By p*) Preferential stripping is used to divide the(, ) rare earths (*are usually divided* ) into two groups, La-Nd and Sm-Gd (the concentration(*tent*) of (*other*) heavier(*y*) rare earths is usually small); (, ) these(*which*) can be further separated into individual rare earth elements if (*necessary*) desired. (*In order*) T (*t*) o reduce reagent(*chemical*) consumption(*s*), (*some*) modified separation process have been (*developed and*) tested at(*in*) pilot plant scale. One approach(*technology is to*) used P204 or P507 to extract Th, F, 【批注】: It isn't obvious why F- should be extracted with a cation exchanger, and since the reference is in Chinese, most readers won't be able to check. Please explain and most of Ce first, then (*and*) the raffinate containing the remaining rare earths (*proceeded to*) underwent further solvent extraction (*process*) steps to(*for*) separate(*ing*) individual rare earth compounds (Figure 11). The (*loaded*) F, Th, and Ce(Ⅳ) (*in th organic phase a*) were selectively stripped from the organic phase. Another (*technology is to*) approach used Cyanex 923 to separate Ce(Ⅳ) from the leachate first. (*and t*) The raffinate containing other rare earths then underwent(*proceeds to a second*) solvent extraction (*circuit*) with N1923 (a primary amine) (*as the extractant* ) to separate Th (Figure 12). Individual rare earth compounds were produced from the Th-free raffinate (*resulted from the second solvent extraction circuit by*) in a third solvent extraction circuit.

*Miscellaneous*

Doyle(, ) et al. developed a novel solvent extraction (*process*) configuration capable of(*that can*) producing(*e*) a mixed Ce-Pr-Nd product (for magnet production) and (*a*) pure Nd oxide simultaneously, with flexibility to alter the relative proportions according to market conditions. The schematic flowsheet for this process is shown in Figure 13. (*The r*) Rare earth chloride solution generated by leaching oxide with HCl underwent (*a*) solvent extraction with P507 in kerosene. In the first solvent extraction circuit, Sm and all heavier rare earths along with Y, were loaded into the organic phase. The raffinate containing Nd and (*all*) lighter rare earths underwent a second solvent extraction (*again*) with P507 in kerosene. Through controlling the number of stages and reflux ratios, Pr and Nd and (*partial*) part of thel Ce were extracted into the organic phase, with the balance of Ce, (*along with*) and all the La (*remained* ) remaining in the aqueous phase from which a marketable (*product of*) lanthanum product were produced. The loaded organic phase underwent selective stripping to produce high-purity neodymium oxide and a mixture of Ce, Pr, and Nd oxides.

Huang, et al. used a synergistic (*extracting*) extraction system to produce different rare earth products from (*the*) rare earth sulfate solutions (*resulted* ) resulting from leaching of bastnesite concentrates (Figure 14). The non-saponified organic phase was used directly to extract rare earths from their sulfate or chloride solutions and by controlling operation conditions, as many as five (or more) commercial rare earth products could be produced simultaneously.

### 3.2 Process simulation

Process development, analysis, control and optimization (*in the*) of rare earth(*s*) solvent extraction (*process are*) is a complex task. (*Computer assisted simulation programs provide useful tools in this area. To develop a useful*) Computer simulation programs for (*to*) monitoring or optimizing(*e*) the rare earth solvent extraction process require(, ) a reliable model for the extraction equilibrium (*has to be defined first*). However, (*though various*) very few models for describing the relevant equilibrium between rare earth elements and different extraction systems have (*been developed, only a few of them*) appeared (*to be available*) in the open literature, and these are usually only (*be*) applicable to a limited and specific range of conditions. This (*is*) probably (*due to*) reflects the similarities of the lanthanides, (*as well as*) their propensity for interactions make it difficult to predict their behaviors in various extraction systems.【批注】: I don't really follow this sentence. It isn't clear why chemical similarity should make it difficult to predict behavior. It is therefore little progress has been made with regard to the development of a general approach for modeling rare earth solvent extraction systems.【批注】: I don't really follow this as you then go on to discuss an approach to modeling rare earth SX.

(*The*) reported (*simulation*) programs for simulating(*on*) rare earth solvent extraction processes (*are*) usually consider(*developed based on the*) countercurrent circuits, due to their ubiquity(*its common use in practice*). S(*The s*) tage-wise calculations(*methods usually preferred due to its higher*) offer efficiency and (*greater*) flexibility. Most (*of the strategies proposed in the literature to perform such simulations*) are based on the McCabe-Thiele method. The technique was originally (*used*) developed for graphical analysis of binary distillation, and later applied to(*has been borrowed for analysis of metal*) liquid-liquid separation processes, especially for (*the one-component*) solvent extraction systems involving (only one extractable species (*is concerned*)). Considering (*the use of*) an n-stage counter-current circuit (*to*) separating(*e*) a metal ion(*element*) from an aqueous solution(*phase*) by solvent extraction, the aqueous stream leaving the $i$th mixer settler would be pumped to the ($i-1$)th mixer settler while the organic phase would be pumped to the ($i+1$)th mixer settler ($i=1, 2, \ldots, n$) (Figure 15). The symbols in Figure (16) 15 are defined as follows: (原文中以下内容为对图片中出现的符号的定义，由于参考价值较小，故略去)

The mass balance (*of the*) for metal ion in the $i$th stage can be expressed as: (公式略)

If the extraction (*equilibriums*) equilibria (*can be obtaine*) dare known, either through theoretical calculations or experimentally, the theoretical number of stage required can be calculated by solving the mass balance equations for all stages if the concentrations of metal ion in the aqueous feed and in the raffinate, and the flow rates (*ratio*) of the organic (*phase to the*) and aqueous phases are known. The concentrations of metal ion in the two phases in different mixer settlers can thus be calculated stage by stage. (*The simplified illustration of*) Figure 16 shows a McCabe-Thiele diagram for (*the operation of*) a 3-stage (*SX*) solvent extraction circuit (*through the McCabe-Thiele diagram is shown in Figure* 16). The equilibrium line OA (*reflecting*) shows the extraction isotherm(*s*) for the desired metal ion(*which usually refer to the variation of metal content in the organic phase with the metal content in the aqueous phase*). The line BC is an "operating line" which is straight with a (*slope*) gradient equal to the ratio of the aqueous (*to solvent*) and organic flow rates ($V_A/V_O$). The points (*in*) on this line reflect the composition of crossing streams in each stage, i. e., ($[M]_{i+1}$, $\overline{[M]}_i$) for all values of $i$ ($i$=1, 2, 3, 4) (*lies on the operating line*). The molar concentration of metal in the organic feed, $\overline{[M]}_0$, is taken (*as*) to be zero in this figure, although it need not be so in practice. When an aqueous feed (with a molar concentration of $[M]_4$) enters (*into*) the 3rd stage mixer-settler, the (*loaded metal in*) composition of the organic phase "crossing" the stream(*where [M]₄ meets the operating line*) will be $\overline{[M]}_3$. T(*hen t*)he composition of the aqueous phase leaving (*the*) stage-3 ($[M]_3$) (*will be determined*) is determined by the point on the equilibrium line with an organic concentration of(*where*) $\overline{[M]}_3$ (specifically, the point ($[M]_3$,) (*meets the equilibrium line*). The (*compositions of metal in the aqueous and*) of the organic stream that crosses the aqueous stream leaving stage 3 is then obtained from the operating line, as shown graphically on Figure 16. This allows calculation of the composition of the raffinate, $[M]_1$. Alternatively, t(*s feeding to and leaving each stage can thus be obtained till the metal concentration of the aqueous phase leaving the first mixer settler ([M]₁) reaches the designed value. T*)he number of stages needed to attain specific stream compositions can (*thus*) be obtained graphically. A similar method can be used to calculate the number of stages needed for the stripping circuit and the corresponding metal concentration in the different streams. An equilibrium line for stripping should be determined or defined first; this will differ from that for extraction because the chemical composition of the aqueous phase will differ, for example, having a different pH in the case of cation or anion exchange extractants, or a different concentration of ligands in the case of a solvating extractant.

A simple case occasionally encountered in rare earth solvent extraction processes is (*the*) a linear equilibrium line (*is a straight line*) through the origin (*metapoint*)【批注】: I'm not familiar with this term to describe the origin, and couldn't find any definition of the term. I'm proposing "origin" as a more commonly accepted term. (OA in Figure 17), (*which indicates*) corresponding to a constant distribution coefficient ($D_M$) for (*the*) metal extraction in all stages. Thus: (, ) (公式略)

Since the operating line, BC, in Figure 18 has a (*slope*) gradient of $V_A/V_O$ as defined in equation (15) 12, (*thus*), then: (公式略)

Assuming $\overline{[M]}_0$=0(no metal in the organic feed) gives: (, *we have,*) (公式略)

Combining equations (17) 14 and (19) 16: (, ) (公式略)

By using similar method Similarly, (*we*) one can (*further*) obtain: (, ) (公式略)

If Defining the(*an*) extraction factor, $E_M$, (*is defined*) as: (, ) (公式略)

E(*quation*) equation (22) 19 can be (*converted to,* ) expressed as: (公式略)

*(or:)*

Thus, (公式略)

Equation ((25) 22) is known as the Kremser equation. *(The)* Its consequences have been discussed *(by many authors)* widely and it *(was)* has been illustrated in various forms. This equation can *(also)* applied*(y)* to any aqueous streams situated an integral number of stages along the mixer settler, not just to streams leaving the mixer-settler circuit. *(It can be also deduced that)* T *(t)* he *(metal concentration in the)* composition of the organic phase leaving any stage can be *(expressed as)* obtained by combining equations 14 and 22: (公式略)

Th*(us, th)* e theoretical number of stages required and the metal concentration in relevant streams can be *(calculated through)* determined from equation (25) 22 if the metal concentration in the feed and *(that in the)* raffinate are defined and the extraction factor is known. *(For)* Similar equations can be developed for stripping if there are*(straight)* linear stripping equilibrium and operating lines (constant stripping factor and constant ratio of aqueous to organic flow rate) *(, equations can be evaluated in a similar manner as for the extraction section to relate the metal concentration in the feed and in the final product)*.

*(However, the extraction of)* Rare earths are seldom extracted in systems containing single rare earths *(elements from an aqueous solution to an organic phase is rarely specific to a single metal only)*. The solvent extraction system chemistry is usually *(chosen so that)* controlled to make the distribution coefficient of the desired element(s) *(is)* much higher than that of the unwanted elements. Even so, *(the)* some extraction of these undesired elements is usually unavoidable *(in most cases)*. In these circumstances, as shown in Figure 4, a scrubbing circuit is commonly employed*(, as shown in Figure 4. This allows the loaded solvent to receive a multi-stage)* to wash impurities from the organic *(with a suitable aqueous solution)* before it leaves*(ing)* the *(extractor)* extraction circuit. *(, which then blends with the aqueous feed passing through the extraction circuit)*. *(The g)* Graphical methods such as the McCabe - Thiele diagram become unmanageable *(when)* for analyzing several extractable species, each of which has their own equilibrium line, and each of which may displace or be displaced by other species in the organic phase *(are present in the solvent extraction circuit)*. In this case, a large number of *(stages of)* extraction, scrubbing, and stripping stages may be required to obtain desired product. *(Some)*

*(a)* Algebraic equations *(based on)* have been developed from the Kremser *(equation)* and mass balance equations to*(have been developed for)* calculate*(ing)* the theoretical number of stages for *(the)* a countercurrent circuit for separating two rare earth elements by solvent extraction, assuming constant distribution coefficients for rare earths in the extraction and scrubbing circuits. *(Constant distribution coefficients have been assumed to rare)* earth element*(s in the extraction or the scrubbing circuits while deducing these equations.)* constant distribution coefficients would not be expected in typical solvent extraction circuits, because the exchange of species between the aqueous and organic phases changes the chemistry of each stream. However, in rare earth circuits where one rare earth is often being exchanged with another*(However, it should be noted that the extraction/stripping behavior of a single rare)* earth element *(will highly depend on the operating*

*conditions. For example, when a cation exchanger is used to extract rare earths from the aqueous solution, the loading of one trivalent rare earth cation results in the release of three protons according to Equation 2. While the aqueous stream from one mixer-settler is delivered to the next stage mixer-settler, the solution pH may have changed in some extent which results in significant deviation between the equilibrium lines for these two stages. However, in some cases, the separation of two rare earths by a solvent extraction circuit is usually the exchange of two rare earth metal ions in each stage* ) ( see equation ( 15 ) 11) , the chemistry may be much more stable, this assumption becomes more realistic. (*and the variation of acidity among different stages is relatively small. Studies also show that the extraction/stripping of the individual rare earths in most cases are nearly independent of the nature of their chemical environment and t*) The separation factors for rare earths are frequently fairly constant throughout a given circuit . As a result, the distribution coefficients of (*all the*) each lanthanide(*s*) can be deduced if the distribution coefficient (*for*) of one has been(*lanthanide is*) determined. Typically, separation factors depend only on the type of selected extractant and to a lesser degree on the anionic species in the aqueous phase. Given that the assumption of constant stream volumes incorporated in the (*Another restriction of the application of*) Kremser equation is (*the constant stream volumes. Most of the organic solvents commonly used for rare earth solvent extraction process do not result in appreciable volume change with the loading of rare*) earth element(*s and the change of the aqueous volume is also insignificant. In these circumstances,* ) usually valid, the equations (*deduced*) derived above can (*usually* ) be (*used to develop*) solved computationally(*er program*) to (*calculate the theoretical stage number required for most rare earth separation processes and to*) analyze and simulate (*the operation of the countercurrent circuit*) rare earth solvent extraction circuits.

Voit developed a simple simulation program for (*an integral circuit for*) producing 99. 5% $Nd_2O_3$ from a (*feed consisting of*) mixed rare earth chloride feed(*s*) ( La, Pr, and Nd) using HEHEHPA (*as the extractant. The separation circuit consisted of anwith extractor*) extraction, (*an*) scrubbing(*er*) and (*a*) stripping sections(*er*) . The Kremser equation was used to (*determine* ) calculate the separation occurring in each section of the circuit. (*By u*) Using same equation, Reddy et al. (*also*) developed a modified simulation program for an integrated rare earth solvent extraction circuit (*for rare earth separation*) . (*The d*) Distribution coefficient data for key rare earth elements (*was*) were tabulated (*as a function of*) for different initial acidity and metal ion concentrations, (*. Then distribution coefficients at any desired acidity and metal ion concentration were determined*) based on operating conditions. (*In case of inconstant distribution coefficient, t*) The use of average separation factors was suggested for non-constant distribution coefficients. (*Some researchers suggested the use of small scale of solvent extraction circuit to verify the calculation before applying the theoretical calculation result in plant scale.* ) Further studies have (*extended* ) considered the(*o*) separation of multi-rare earth elements with two or more outlets/products (*by one*) in a single countercurrent circuit, . (*The combination of the*) A simulation system has been combined with(*and the*) on-line EDXRF (*analytical technique has been used*) analysis to monitor the steady or dynamic performance of stage-wise processes. The latter authors(*y*) claimed that their model was especially useful when there was no more than one intermediate feed point and the distribution ratio of the component involved was constant. The detail of the calculation method followed was not reported.

## 4 Summary

*(R)* are earth elements have unique properties and are becoming *(uniquely and )* critical in many high-technology applications*(industry)*. China is currently the world's largest producer of rare earth elements providing more than 95% of the world's total supply. *(A variety)* Various *(of)* rare earth minerals have been identified, of which *(and)* bastnesite (La, Ce) $FCO_3$, monazite, (Ce, La, Y, Th) $PO_4$, and xenotime, $YPO_4$, are the most commercially important *(resources of rare earths)*. Rare earth minerals are usually concentrated by flotation or gravity methods to produce concentrates that *(can be)* undergo *(fed to subsequent)* hydrometallurgical processing to recover rare earth metals or compounds. Rare earth concentrates or calcine *(is)* are usually leached with an inorganic acid, such as HCl, $H_2SO_4$, or $HNO_3$.

*(After solution purification, separation processes based on solvent extraction techniques are used to yield)* Individual rare earths or mixed rare earth solutions or compounds. Rare earth producers usually follow almost identical principles or schemes when *(on)* selecting *(process routes for)* solvent extraction circuits to separate*(ion of)* rare earths from *(a)* each other. Usually *(aqueous solution, which includes separation of rare earth elements in their)* trivalent *(states, separation of)* rare earths are separated into two or more groups, followed by subsequent separation of individual rare earths and preferential separation of yttrium if possible. *(Rare earth solvent extraction processes can be roughly identified as two classes, primary separation which aims to separate rare earth elements from other elements, and secondary separation which aims to separate single or mixed rare earths from the mixed rare earth product obtained from primary separation.)* The choice of extractants and aqueous solution conditions is influenced *(by)* both by cost considerations and technical requirements *(of technical performance)*, such as selectivity. The use of cation exchangers, solvation extractants, and anion exchangers, for separating rare earths has been extensively studied. Commercially, D2EHPA, HEHEHP, Versatic 10, TBP, and Aliquat 336 have been widely used in commercial rare earth solvent extraction *(practice)*.

On the aspect of *(process and equipment, u)* Up to hundreds of stages of mixers and settlers may be assembled together to provide mixing and settlingseparate all the individual rare earths in a feedstock. Typical configurations for rare earth solvent extraction circuit have been reviewed. *(Much effort has been made on computer simulation of the solvent extraction circuit for separating rare earths, however, little progress in the development of rare earth simulation programs has been reported, which is probably because there are many variables to be considered due to the co-occurrence of many rare earths in the leachate which result in complicated equilibriums of extraction/stripping. The )* T*(tr)* aditional graphical methods for simulating solvent extraction circuits, such as McCabe-Thiele diagrams, *(usually can not be directly applied for)* have limited utility for rare earth solvent extraction circuits. However, more promising computational approaches based *( Some methods of calculating the theoretical number of stages based )* on the Kremser *(equation)* and mass balance equations have been developed.

# 8.2 范例二

## Mercury recovery from contaminated solid waste: A mini-review

Feng Xie*, Kaiwei Dong, Wei Wang

School of Metallurgy, Northeastern University, 3-11 Wenhua Road, Shenyang, P. R. China 110004

**Abstract**

Mercury (*is*) continues to receive(*iving more concerns due*) attention across a variety of disciplines due to its high mobility and high toxicity to human health and the environment. Sources of mercury -containing solid waste may come from nonferrous metallurgy, mercury mining and the chemical manufacturing industry. (*Many*) Significant efforts and resources have been (*undertaken*) leveraged to develop effective remediation technologies to reduce the hazardous effect of mercury contaminated waste. Research and development on mercury recovery from mercury contaminated wastes (*has been*) is reviewed in this paper. Though the thermal treatment(*ing*) technologies(*y*) (*is commonly*) are used in practice, leaching processes including acid leaching, alkaline leaching and bioleaching, using different lixiviants, have (*been*) also been developed. It i(*seem*) s difficult to (*considerer*) conclude which (*is nowadays the best*) technology (*to be applied*) offers the best outcomes. However, leaching processes can (*be usually*) normally be performed at the site of contamination and can usually remove mercury permanently from wastes in a relatively safe and controlled (*under controlled/properly-designed operating procedures compared to the thermal technology*) manner - this is not always true for thermal treatments. These leaching processes can be used independently or in conjunction with other treatment technologies. The selection of leaching (*lixivants*) lixiviants is highly dependent(*ing*) on the characteristics of the different waste streams. Speciation and coordination of mercury in waste (*sometimes are*) can be critical for application of a recovery technology, especially for (*those*) leaching processes.

**Keywords:** Mercury, Recovery, Solid Waste, Leaching.

## 1 Introduction

Mercury is receiving more concerns due to its high (*toxicity to human health and environment.*) It is well established from t(*T*) oxicology studies (*proved* ) that mercury, both inorganic and organic, can cause serious health effects. O (*though the o*) rganic mercury is, however, generally more toxic (Miretzky and Cirelli, 2009 , UNEP, 2013 , U. S. EPA, 1997a). Mercury contamination can also be much more widespread than that of other heavy metals due to the high mobility of mercury (Fitzgerald et al., 1998 , Selin, 2009 , Svensson, 2006). Mercury may undergo complex physical, chemi-

* Corresponding author. Tel.: +86 24 8367 2298; .E-mail address: xief@ smm.neu.edu.cn

cal and biological transformations in the environment, e. g. , the atmospheric transport of Hg, the photochemical oxidation and subsequent deposition of mercury on water and land and (, *and* ) the methylation of $Hg^{+}$ by reducing bacteria in anoxic habitats. The subsequent (*and its* ) absorption and accumulation by organisms (*which*) results in high mercury concentrations in fish and chronic low level(*r*) exposure of humans through the food chain ( Kudo and Miyahara, 1991 , Mason et al. , 1996 , Wagner-Dobler, 2003) . (*Thus*) Thus, mercury containing waste is regarded as a dangerous substance, which in most cases is classified as a hazardous material. In order to (*strengthen actions on mercury control to*) reduce mercury pollution worldwide, in February 2009, at the 25th Session of the United Nations Environment Program ( UNEP) Governing Council, in Nairobi, Kenya, more than 140 other countries agreed to begin negotiations on a legally-binding instrument to control the shipment of mercury. This (*has*) instrument was(*been*) successfully developed into the Minamata Convention on Mercury (*later*) ( UNEP 2013) . As a result, (*policies on*) mercury control policies are being strengthened by the countries involved in the treaty and the application of mercury recovery and detoxification technologies is encouraged. (*especially mercury recovery and stabilization/detoxification is always encouraged.* ) However, mercury recovery from waste is often discouraged, mainly for economic reasons. It was reported that recovery units have been established in countries like Germany, France, Austria, and Sweden, 【批注】: References needed but the total amount of secondary mercury recovered from waste (*was*) is not known. In this work, the research and development on methods and technologies for(*of*) mercury recovery from solid waste and some stabilization or immobilization technologies (*has been*) are reviewed.

## 2 Sources of mercury-containing solid waste

The e(*E*) stimation of the total quantity of mercury in waste worldwide is (*a*) difficult (*work*) and (*still*) remains incomplete ( Chen et al. , 2016 , Li et al. , 2017) . Telmer and Veiga ( Telmer and Veiga, 2009) estimated that approximately 1000 metric tons of Hg per year was released from the artisanal and small scale gold mining industry in at least 70 countries worldwide. Wang et al. ( Wang et al. , 2012) estimated (*of*) that global mercury emissions ranges from 5000 to 8000 metric tons per year. Mercury emission may come from natural, anthropogenic and re - emitted resources. Typical natural mercury emissions include volcanic emission, forest fires, and degassing from mercury mineral deposits. Traditionally, mercury-contaminated solid waste mainly results from mercury mining and (/) mineral processing, non - ferrous metallurgical processing, the chemicals manufacturing industry, and municipal and medical wastes. The e(*E*) mission of mercury from coal and oil combustion and other mercury-contaminated resources may also resulted in local environmental contamination ( Haiyan and Stuanes, 2003) . 【批注】: Only local? I remember reading articles about long range transport from power plants and roasters. In fact, I think one of the Barrick roasters in Australia released Hg that was detected in South America.

Mercury mining activities usually produce large quantities of abandoned mine-wastes. A typical example is the mercury-laden mining tails produced in the area of Xunyang, Shaanxi Province, in which the mercury (*content*) concentration varies from several hundred to several thousand( *s*) ppm. Mercury may be released into the local environment through erosion of these mine wastes. It has been demonstrated that soluble mercury can leach from (*calcines* ( ) tailings ( ) *residual*) after small-scale refining of the metal (Ping et al., 2008). Soil close to these mercury mining districts has been heavily polluted due to the extensive mercury mining/refining in these areas. Elevated methyl mercury levels in soil have been found in rice paddy fields near the Wanshan and Wuchuan mercury mines of China (methyl mercury concentrations in soils were 23 and 20μg/kg,【批注】: If you use ppm above maybe you should use it everywhere? respectively) (Qiu et al., 2006).

(原文中此处修改内容较少，且类似问题在其他篇幅出现过，故在此略去部分篇幅)

## 3 Mercury recovery

【批注】: Not clear here what you mean by recovery – does extraction alone represent "recovery"? Usually recovery implies extraction + precipitation in some form...

Many efforts have been undertaken to develop effective remediation technologies to (*reduce* ) mitigate the hazardous effects of mercury contaminated waste to the environment and to human health. Various remediation techniques have been reported for disposing of mercury-containing wastes and these methods can be generally classified as stabilization/solidification (including vitrification and chemical formation) and extraction (including thermal treatment and extraction with chemicals) (Randall and Chattopadhyay, 2013, Wang et al., 2012). The selection of mercury recovery techniques is highly dependent (*ing*) on the characteristics of the different waste streams. The total mercury concentration in waste is an important factor, but it would be insufficient (*only on this*) to justify a (*suitable*) recovery method based only on this metric. In most cases, speciation and coordination of mercury in waste (*sometimes are*) is critical (*for application of a*) the choice of a recovery technology(*, especially for those extraction processes*).

### 3.1 Chemistry of mercury in waste

(原文中此处修改内容较少，且类似问题在其他篇幅出现过，故在此略去部分篇幅)

### 3.2 Mercury extraction technologies

Extraction of mercury from mercury-contaminated wastes is always favored during detoxification. Thermal treatment and water-based technologies (or leaching technologies) that use both physical and chemical separation methods to reduce contaminant concentrations in wastes have also commonly been used (Dermont et al., 2008b, U.S. EPA, 2007a). Typical technologies for (*of*) mercury extraction from mercury contaminated waste are summarized (*at*) below.

#### 3.2.1 Thermal treatment

Mercury (*has a potential to become*) can vaporize (*in the*) at【批注】: Do you mean that it has a high vapour pressure? room temperature (*so*) such that (*the*) thermal treatment (*could*) may (*been*) used efficiently. (*The t*) Thermal treatment using a rotary solar kiln【批注】: Specify in parentheses what this is to treat *(the)* mercury mine wastes has been reported (Navarro et al., 2014). The comparative*(ble)* results *(has)* showed*(n)* that a rotary kiln is more efficient than a fluidized-bed reactor. *(The)* An off-gas treatment system was also used during the desorption process to control *(the air)* emissions to the air (ITRC, 1998). Laing et al. (Laing et al., 2013) investigated the effect of heating temperature, duration and particle size of the polluted materials, and thickness of the polluted material layer on the efficiency of Hg removal from two polluted soils collected from Belgium and mining residues from Nicaragua. The Hg removal efficiency was found to increase with increasing temperature. The effect of particle size on Hg removal efficiency was dependent*(ing)* on the treatment temperature. When temperature was set between 100 and 250℃, the extent of removal *(efficiency)* varied*(s)* with *(respect to)* particle size (more efficient removal with smaller particles). However, when the temperature wa*(i)*s set above 300℃, the extent of Hg removal *(efficiencies are)* was no longer affected by particle size. The extent of removal *(efficiency)* when treating the <1mm fraction of the most polluted soil for 72 hours at 200℃ was found to decrease from about 90% to around 80% when the thickness【批注】: Thickness? Do you mean particle size? of the soil layer increased from <0.5cm to around 3cm. Massacci et al. (Massacci et al., 2000) examined the combination *(method)* of screening, attrition and thermal treatment to remove mercury from mercury-polluted soils. The finest particles from screening and attrition of contaminated concrete contained 34% of the mercury in the sample, at a grade of about 8500 ppm. Experimental results indicated that thermal treatment was a feasible method to remove mercury from contaminated soils. Ma et al. (Ma et al., 2015) developed a novel method for disposal of a mercury contaminated soil-in which citric acid was used to facilitate the thermal treatment process. At the optimum molar ratio of citric acid to Hg, the mercury concentration in soils was successfully reduced from 134mg/kg to 1.1mg/kg when treated at 400℃ for 60 minutes. The treated soil retained most of its original physiochemical properties. It was found that during the treatment process, citric acid could provide an acidic environment which enhanced the volatilization of mercury. The thermal treatment method facilitated by citric acid also reduces *(35%)* the required energy input by 35% as compared to the traditional thermal treatment method. Wang et al. (Wang et al., 2016a) studied the thermal treatment of coal fly ash which contained a significant amount of mercury.【批注】: I think you should be more quantitative here and throughout. How much Hg did this contain? Mercury in the ash sample began to *(emit)* evaporate at approximately 200℃ and about 20% of the total mercury *(could emit)* was recovered at 206℃. The emission rate of mercury increased sharply when temperature was higher than 206℃ and it was almost completely *(emitted)* evaporated from the ash at 1200℃. Busto et al. (Busto et al., 2011) have studied the thermal treatment of a mercury contaminated soil and close to 100% mercury removal *(efficiency)* was obtained.【批注】: Under what conditions?

### 3.2.2 Leaching

Leaching is an important process for metal extraction from various materials (Xiu et al., 2013). Generally, the target metal is first dissolved into solution through *(reacting)* reaction with the *(leac-*

*hing)* lixiviant and is subsequently *(recovered )* separated from the leachate for final recovery *(through separation)*. This separation *(, )* may include various steps such as solution purification, *(and recovery processes)* precipitation and/or electro-reduction processes. *(While)* *(The separation methods for)* of mercury *(recovery)* from aqueous solutions may *(including)* include cementation, adsorption, ion exchange, solvent extraction, precipitation, and membrane filtration. However, effective *(have been well documented in literature, )* leaching is always most critical in hydrometallurgical processes (Yang et al., 2006). Compared to other detoxification/stabilization technologies, the advantages of leaching technologies include: (1) the *(treating)* processes can *(be usually)* normally be performed at the site of contamination, avoiding risks associated with transporting the contaminated waste off-site to a treatment facility; (2) the processes usually quantitatively remove mercury *(permanently)* from wastes and allow further mercury recycling *(and)*. T *(t)* he processed solid, if no *(presence of )* other hazardous components are present, can then be return to- or stockpiled at *(the)* site, e. g., landfill soil; (3) the processes are relatively safe under controlled/properly-designed operating *(procedures)* conditions and they avoid Hg evaporation to the local environment. Leaching processes can be used independently or in conjunction with other treatment technologies. Different lixiviants have been used for extracting mercury from contaminated soil. W*(, of which w)* ashing with water is usually conducted first (Abumaizar and Smith, 1999, Dermont et al., 2008a, Reis et al., 2014, SEPA, 2009, Sierra et al., 2010, U. S. EPA, 2007b, Wang et al., 2012, Xu et al., 2014). Figure 3 presents a schematic of a bench-scale soil washing process, including particle size classification and oxidative leaching, demonstrated by Cognis, Inc. of California on two different batches of highly mercury contaminated soil (D. J. Stephan, 1993, D. J. Stephan et al., 1995). Reis et al. (Reis et al., 2014) reported an optimization study on the extraction of the water-soluble mercury fraction from soils. Test results indicated that the ratio of soil and water does not have a significant influence on the extraction of water-soluble mercury, although it is advisable to keep the ratio as low as possible to guarantee that all soil contacts the water. *(The k)* Kinetic studies by Reis et al. (2014) *(y)* showed that it takes up to 24 h for extraction to reach equilibrium, and that the mercury removal reaction takes place in two stages, a faster *(one)* stage followed by a slower stage, which can be described by the following two first-order reaction model: (公式略)

where $Q_1$ and $Q_2$ (mg kg$^{-}$) are the mercury concentration extracted in the first and second stages, respectively, and $k_1$ and $k_2$ 【批注】: Can you add the exact numbers here for rate constants in parentheses? are the apparent rate constants. This*(e)* reaction model *(reveals that some water-soluble Hg species are extracted more quickly the second stage as $k_1$ is always larger than $k_2$, )* suggest*(ing)* s that *(they)* various mercury species *(may)* are *(be)* bound differently to the matrix. For example, water-soluble mercury species in clay soil may be extracted preferably in the second stage while in sandy soils, the water-soluble mercury species may be extracted mainly in the first stage. However, since various mercury compounds frequently occur in contaminated wastes, leaching simply with water can normally only dissolve *(partial)* mercury in ionic form*(s in most cases)*. Processes of recovering mercury from contaminated waste with or without oxidative and complex reagents in acid and alkaline media have thus been developed.(图略)

3.2.2.1 *Acid leaching*

Because divalent mercury reacts easily with chlorides to form salts that are soluble in water, such as $HgCl_2$ (Bilinski et al., 1981, Carpi, 1997), HCl is commonly used for leaching mercury contaminated waste. According to the Eh-pH diagram of the Hg-Cl-$H_2O$ system (Figure 4), $HgCl_2$ is stable at lower pH and complexes of mercury and chloride may form at higher ratios of $Cl^-$ to $Hg^+$ in solution (Figure 5). However, it should be noted that some mercury species such as HgS cannot be effectively dissolved in HCl solution even at elevated temperatures because of its low solubility in HCl (Dronen et al., 2004). Fernandez-Martinez and Rucandio (Fernández-Martínez and Rucandio, 2005) examined the extractability of different mercury species in two acid solutions, 50% v/v HCl and 50% v/v $HNO_3$ and the results are shown in Table 1. According to Table 1, most *(tested)* of the mercury species were quantitatively dissolved in both acids with the exception of HgS and $Hg_2Cl_2$. While $Hg_2Cl_2$ is not usually found in the environment, HgS represents a common source of mercury in soils, especially those from mercury mining areas (Egler et al., 2006). The application of both acid solutions to leach HgS produced low solubility of mercury: less than 5% for【批注】: This is not a solubility - can you add units? both 50% v/v $HNO_3$ and 50% v/v HCl. Revis et al. (Revis et al., 1989) reported that HgS in soil could not be effectively extracted even *(by using)* with strong (12mol/L) $HNO_3$ solution. However, the presence of *(some)* other compounds may significantly increase the dissolution of HgS in these acid solutions. For example, i*(I)*n the case of HCl, the *(extractability)* extraction of HgS is enhanced by the presence of KI, $CuSO_4$, $MnO_2$ and $NaNO_3$. Chloride and iodide compounds are also able to promote the *(leachability)* leaching of HgS when a nitric acid solution is applied. In the presence of $MnO_2$ and $NaNO_3$, complete extraction of Hg is achieved when HCl solution is used to leach mercury contaminated soils from cinnabar mines (Fernández-Martínez and Rucandio, 2003, Fernández-Martínez and Rucandio, 2005). (图、表略)

A hydrometallurgical process for recovering mercury from cinnabar ore by leaching with a hydrochloric acid- potassium iodide solution has been reported (Núñez and Espiell, 1984). The leaching experiments were performed at *(stirring)* mixing speeds fast enough to eliminate the effect of this variable in the overall reaction rate, and the $H_2S$ generated in the reaction was removed from the reaction vessel by an air flow. The process consisted basically of three successive stages. Cinnabar was leached with the hydrochloric acid-potassium iodide solution and the resultant $H_2S$ was collected*(, and the $H_2S$ generated in the reaction was removed from the reaction vessel by an air flow)*. The leaching stage was then followed by electrolytic treatment of the acid tetra-iodo-mercuriate solution, thus obtaining very pure metallic mercury on the cathode and iodine *(on)* at the anode. The iodine remains in solution if an excess of iodide ions was present as the insoluble $I_{3-}$ complex.【批注】: Complex of what iodine and? The spent electrolyte was treated with the $H_2S$ produced in the leaching reaction to regenerate the iodide ions and the acid. Thus, in the regenerated solution all the reagents had the same concentration as at the beginning and could be recycled. Under experimental conditions, the dissolution rate of mercury appeared to be controlled by chemical reaction on the cinnabar surface. The dissolution rate was of the first order with respect to hydrochloric acid【批注】: It would be nice here to write the reactions for each stage activity and of second order with respect to potassium iodide activity.

Núñez-Núñez et al. (Núñez et al., 1986) also reported a process of leaching cinnabar ore with HCl-thiourea solutions. (*The t*) Thiourea (*favors the* ) coordinat(*ion*) es with mercury to form $HgCl_2 * 2SC(NH_2)_2$, the solubility of—which (*solubility d*) depends on temperature and HCl and thiourea concentrations (IUPAC, 1979). (*It*) This complex may be precipitated from leaching solutions by cooling or addition of HCl gas if necessary. The dissolution rates of mercury were determined over a temperature range of 60 to 100℃, at an acid concentration of 0 to 5M and thiourea concentration ranging from 0.3 to 1.5M at atmospheric pressure. The experimental results indicated that the dissolution rate of mercury was (*of the*) second order (*for*) with respect to the thiourea concentration, first order for HCl in the concentration range of 0 to 2M, and zero order for HCl concentrations above 3M. An activation energy of 53.6kJ/mole was found. To recover mercury from the leachate, it could be precipitated as $HgCl_2 * 2SC(NH_2)_2$ by cooling or addition of HCl gas if necessary. The leaching solution could also be treated electrolytically, with metallic mercury depositing on the cathode and thiourea oxidizing on the anode to formamidine disulfide. The spent electrolyte was recycled to the leaching reactor and formamidine disulfide was reduced to thiourea by $H_2S$ generated in the cinnabar attack. Ballester et al. (Ballester et al., 1988) developed a hydrometallurgical process for cinnabar treatment which deals with an ore using a hydrobromic acid-like reagent in the form of tetrabromomercuriate ion ($HgBr_4^-$). The best conditions to leach cinnabar were 303 or 313K and 7 kmol· $m^-$【批注】: Again, since this is a review, I suggest rationalizing all of the units throughout the manuscript. HBr.

Hintelmann and Nguyen (Hintelmann and Nguyen, 2005) developed an efficient extraction method based on acid leaching ($HNO_3$) with the purpose (*to*) of (*measure*) measuring the methylmercury (MeHg) in benthic organisms and plant material. The digestion process used 5 mL of 4mol/L $HNO_3$ at 55℃ to leach MeHg (*of*) from 20mg of tissue and plant material. The acid digestion resulted in 96% ± 7% recovery of MeHg from oyster tissue and 93% ± 7% from pine needles. Jang et al. (Jang et al., 2005) applied a rotary shaking process using different acid solutions to extract mercury from fluorescent lamps. For both acid solutions (HCl and $HNO_3$), the extracted mercury fraction seemed to increase with an increase in acid concentration. The mixture of nitric and hydrochloric acid solution exhibited higher mercury extraction from the glass than the nitric acid solution only. The extracted mercury fraction increased sharply to about 35% when 5% of the mixed acid solution was used. When the mixed acid concentration was higher than 5% (*of*), however, the mercury extraction exhibited insignificant change. The maximum extracted mercury fractions were 36% for the mixed acid and 28% for the nitric acid only. Rey-Raap and Gallardo (Rey-Raap and Gallardo, 2013) also examined the feasibility of using mixtures of HCl and $HNO_3$ to remove mercury (*bonded*) bound in residual glass from spent fluorescent lamps. The experimental data (*indicated*) showed that (*by using the*) acid solutions were much more effective than water for mercury extraction(*, the percentage of mercury extracted is much higher than by using ultrapure* water.) Under optimized conditions, more than 69% mercury was removed from the waste glass. (*They*) These authors believed that because HCl reacted easily with divalent mercury, (*and*) the acid solution not only removed the mercury in the phosphorous powder attached to the surface of the glass but also the mercury that had diffused through the glass matrix (Rey-Raap and Gallardo, 2013). Al-Ghouti et al. (Al-Ghouti et al., 2016) investigated mercury removal from a phosphorous powder by mixtures of nitric and hydrochloric acid at ambient conditions with and without microwave-assisted leaching. They found that increasing the concentration of the nitric acid and using microwave-assisted leaching could improve the (*efficiency*) extraction significantly. Test results showed that the microwave-assisted leaching (*was*) almost doubled the amount of mercury leaching compared to acid leaching only, leading to a maximum mercury leaching efficiency of 76.4%. Zhou and Dreisinger (Zhou and Dreisinger, 2017) reported a process using hypochlorite to leach elemental mercury. The process (*consists*) consisted of extraction of elemental mercury into solution to form aqueous mercury (Ⅱ) and mercury precipitation as mercury sulfide or mercury selenide. It was found that elemental mercury could be effectively extracted by using hypochlorite solution in acidic environment to form aqueous mercury (Ⅱ) chloride. The effect of different parameters on the extent and rate of mercury extraction were studied and test results showed that near complete extraction could be achieved within 8 hours by using excess sodium hypochlorite at pH 4 with a (*fast stirring*) mixing speed of 1000 rpm.

3.2.2.2 *Bioleaching*

Bioleaching is (*one of*) a common leaching methods in hydrometallurgy (Boon and Heijnen, 1998). It (*favors*) is most commonly applied for leaching (*of the*) low (*concentrate*) grade ores, which it does at (*with*) relatively low cost. Bioleaching (*though it*) also has disadvantages, such as long leaching times and a dependence on (*sensitive b*) bacteria (*l condition*), which require careful control of solution conditions. (*which s*) Sometimes these factors limit its application. Bacteria such as *acidithiobacillus ferrooxidans and acidithiobacillus thiooxidans* have the ability to oxidize iron and sulfur and thus promote metal dissolution (Rohwerder et al., 2003). Bioleaching of sulfides with these two bacteria is believed to occur via two steps: the chemical oxidation of the sulfide by Fe(Ⅲ)$^{+}$ and the bacterial regeneration of Fe(Ⅲ) from Fe(Ⅱ)$^{+}$ (*as leaching agent*). Wang et al. (Wang et al., 2013a, Wang et al., 2013b) examined the feasibility of bioleaching of cinnabar and the effects of temperature, initial pH of the solution, dilution rate and iron on the mercury dissolution. The reactions involved(*s*) can be expressed as (*following*) follows: (公式略)

Ferrous iron can be oxidized to ferric iron rapidly by A. ferrooxidans: (公式略)

The results demonstrated cinnabar dissolution had a strong relationship to the bioprocess of A. ferrooxidans and the iron concentration (*tightly*). Though the increase of ferric iron in solution would facilitate the (*dissolving*) dissolution of cinnabar, it would also facilitate jarosite and Schwertmannite precipitation at increasing pH through the following reaction: (公式略)

where $M^{+}$ represents monovalent cations such as $NH_4^{+}$, $K^{+}$, $Na^{+}$ and $H_3O^{+}$.【批注】: I think somewhere in this section you should discuss the toxicity of Hg to bioleaching bacteria – I think it is like Ag and I think Hg kills the bacteria. The precipitates adsorbed on the surface of cinnabar particles would have a negative effect on the transport of dissolving mercury from the substrate because of kinetic barriers (Bernaus et al., 2006). Under optimized conditions, a maximum mercury concentration of 1.38g/L was【批注】: What was the percent extracted? achieved after bioleaching 5.0g cinnabar with 100mL culture solution for 15 days. Dronen et al. (Dronen et al., 2004) conducted an assessment of acid wash and bioleaching pre-treating options to remove mercury from coal. They reported that mercury removal from exposure to bacteria in the culture solution was only slightly higher than that from the control samples, which is most likely due to the solvent effect of the acidic culture media rather than any metabolic reaction by the bacteria. It seemed that the bioleaching technique did not appear to be effective for the removal of mercury from (*this*) the lignite coal they studied.

3.2.2.3 *Alkaline leaching*

Alkaline leaching is also used to extract mercury from contaminated solid. Thiosulfate solutions (*is suggested as the leaching reagent*) have been studied for the extraction of mercury from mercury-containing waste by many researchers (Issaro et al., 2010, Ray and Selvakumar, 2010). The Eh-pH diagram of meta-stable $Hg-S-H_2O$ system and the speciation diagram of $Hg-S_2O_3^-$ under different pH are shown in Figure 6 and Figure 7, respectively. Han et al. (Han et al., 2017) found that mercuric oxide can react quickly and (*thoroughly*) quantitatively with thiosulfate to form the mercuric thiosulfate complex. The dissolution reaction can be expressed as following: (公式略)

Test results indicated that the solution pH (*exhibits*) had a significant effect (*of*) on HgO leaching when the molar ratio of $Na_2S_2O_3$ to total mercury (*is*) was relatively low (e.g., <3). An increase of $Na_2S_2O_3$ concentration (*may potentially*) increased the dissolution of HgO due to the increase of molar ratio of $Na_2S_2O_3$ to mercury. Increasing temperature (*can*) slightly (*increases*) increased the dissolution rate of HgO in $Na_2S_2O_3$ solutions when the temperature (*varies*) varied from 293K to 323K. The apparent (*active*) activation energy (*can be calculated according to the slope of the line, giving a value of*) was calculated to be 23.3kJ/mol and the initial leaching of HgO in $Na_2S_2O_3$ solution (*is*) was under mixed control. Under optimum experimental conditions, virtually all mercury oxide can dissolve in $Na_2S_2O_3$ solutions within five minutes, indicating the thiosulfate leaching (*system*) is an effective way for detoxifying (*those*) mercury solid wastes containing mercury oxide. Lu et al. (Lu et al., 2014) examined the effect of thiosulfate on mercury removal and reported that $Na_2S_2O_3$ can be used as a chemical aid for improving trace mercury removal (Dyrssen and Wedborg, 1991). However, it was reported that only 4% of mercury can be removed from the contaminated soil while leaching with simple $Na_2S_2O_3$ solutions【批注】: At what pH? (0.01mol/L) (Stumm and Morgan, 1996). Issaro et al. (Issaro et al., 2010) also reported that only 50±5% of mercury can be extracted from the contaminated soil after leaching with 0.01M $Na_2S_2O_3$ for about 24 hours.【批注】: Did they specify the type of mercury in soil? If yes, then say what it was. If not, then say that they didn't. This is probably due to the different occurrence of mercury in these soil samples. Han et al. (Han, et al., 2017) reported that under simi-

lar leaching conditions, elemental mercury, mercury sulfide and some organic mercury (*virtually do*) did not dissolve in simple $Na_2S_2O_3$ solutions. Issero et al. (Issaro et al., 2010) suggested the use of $H_{2(2)}O_{2(2)}$ to pre-oxidize these species before thiosulfate leaching and ideally, the oxidization reactions can be expressed as (*following*) follows: (公式略)
where $R$=organic radicals such as $CH_3$, $C_2H_5$, $C_6H_5$. The formed $HgO/HgSO_{4(4)}$ can thus be easily leached from soil with ($Na_2S_2O_3$) -$Na_2S_2O_3$ solutions. However, the effect of pre-oxidization with $H_{2(2)}O_{2(2)}$ (*is*) was insignificant on subsequent mercury extraction with thiosulfate while dealing with a mercury contaminated soil (Han et al., 2017). Further (*study*) studies on the leaching behavior of mercuric sulfide with cuprous-thiosulfate solutions (*has been*) were conducted.【批注】: By who? Test results indicated that initial pH, $[Cu]/[S_2O_3^{2-}]$ ratio, and temperature may exhibit a significant effect on both the stability of cuprous-thiosulfate solutions and the mercury leaching. Increasing temperature can increase mercury leaching from HgS when the temperature varies from 293K to 313K. Under optimized experimental conditions: initial pH 8.50, molar ratio of copper to thiosulfate 0.1 and leaching temperature 313K, the mercury (*leaching rate*) extraction is 92.37% after 7-hours of leaching. (*The study*) Studies on the leaching mechanism indicated that $Cu^+$ can be substituted by $Hg^+$ to form soluble $Hg(S_2O_3)_n^{n-}$ in solution and $Cu_2S$ precipitates on the surface of unreacted HgS particles. Though the kinetics of HgS leaching with cuprous-thiosulfate solutions fitted well to the shrinking core model with diffusion control, the calculation of the activation energy indicates the leaching process may be mix controlled by diffusion and surface chemical reaction. (*They*) These authors suggested that cuprous-thiosulfate solutions can be potentially used for mercury leaching from HgS contaminated solid waste.

Mixtures of NaOH and $Na_2S$ can be used to extract mercury from mercury sulfide (*mineral*) according to the following reaction:【批注】: Are you missing a reference here? Do you have the K for this reaction?(公式略)
Feng reported a process using NaOH+$Na_2S$ solution to leach (*the*) mercury mine tails, which contained both Hg and Sb. Under optimized conditions: 0.5mol/L NaOH, 2mol/L $Na_2S$, liquid to solid ratio of 5 : 1 (mL/g), 65℃(*℃*) and after 2 hours leaching, extraction of Hg and Sb (*can*) reached 93% and 54%, respectively. Wang et al. (Wang et al., 2007) conducted a study on mercury leaching with ammonia from fly ash. Based on the formation constants of mercury-hydroxide, mercury-ammonia complexes, and the (*acidity*) dissociation constant of the ammonium ion, the mercury speciation as a function of pH in【批注】: It would be good to add these constants in a Table - especially for the Hg species the presence of $10^-$ mol/L ammonia was calculated (Figure 8). (*It*) Figure 8 shows that ionic mercury ($Hg^+$) and free ammonia ($NH_3$) can form mercury-ammonia complexes (*in*) over a wide range of pH (between 2 and 10.5). When pH is greater than 10, the fraction of $Hg(OH)_2$ increases significantly. Because $Hg(OH)_2$ has high affinity for many sorbents, high concentrations of ammonia enhanced the mercury leaching in the alkaline solution due to the formation of less adsorbable mercury-ammonia complexes.

Ozgur et al. (Ozgur et al., 2016) investigated an oxidative leaching and electrowinning processes for recovering mercury from spent tubular fluorescent lamps. They found mercury could be leached with (*an efficiency of*) up to 96% using 0.5M/0.2M NaOCl/NaCl (*reagents*) at 50℃ (*C*) and pH 7.5 over 2 hours(*for 2-h*). The e(*E*) lectrowinning process was conducted (*on*) from the filtered leaching solutions and over the 81% of the mercury was recovered at (*the*) a graphite electrode using citric acid as a reducing agent.【批注】: Why was this used during electrowinning? I think this needs more explanation. The results indicated that the process of oxidative leaching with NaOCl/NaCl, followed by electrowinning, appears to be technically feasible for mercury recovery from spent fluorescent lamps. (图略)

## 4 Summary

Mercury is receiving more concerns due to its high mobility and high toxicity to human health and the environment. Sources of mercury containing solid waste may come from mercury mining, non-ferrous metallurgy and the chemical manufacturing (*industry*) industries. Many efforts have been undertaken to develop effective remediation technologies to reduce the hazardous effect of mercury contaminated waste to environment and to human health. Various recovery technology including thermal treating and leaching have been reported for disposing of mercury - containing wastes. Processes of acidic leaching, alkaline leaching and bioleaching using different (*lixivants*) lixiviants ($HCl$, $HNO_3$, $NaClO_3$, $Na_2S_2O_3$, and $Na_2S$, etc.) have been developed. These processes can be used independently or in conjunction with other treatment technologies. (*It is difficult to considerer which is nowadays the best technology to be applied, because of the different chemical and physical characteristics and properties of the wastes to be treated. However, a*) An obvious advantage of the leaching technologies is that they can (*be usually*) normally be performed at the site of contamination, avoiding risks associated with transporting the contaminated waste off - site to a treatment facility. Another advantage is that they usually quantitively remove mercury (*permanently*) from wastes and allow further mercury recycling in a relatively safe (*under controlled/properly-designed operating procedures compared to the thermal technology*) manner.

# 8.3 范例三

## Research on (*Leaching*) leaching (*gold from*) of carbonaceous gold ore with copper-ammonia-thiosulfate solutions

Jian Wang, (*Yan Fu, Feng Xie*) Wei Wang*, (*Wei Wang*), Wenshuang Xu, Yan Fu, Feng Xie

School of Metallurgy (*Engineering*), Northeastern University,
#3-11(,) Wenhua Road, Shenyang, (110004,) PR China(.) 110004,

**Abstract**(*:*)

For (*the*) those carbonaceous gold ore or concentrate, gold extraction (*is usually low*) by direct cyanidation is usually low due to the known preg-robbing effect. In this paper, (*gold*) leaching of a carbonaceous gold ore containing high content of dolomite with copper – ammonia – thiosulfate solution (*from carbonaceous gold concentrate containing dolomite*) has been investigated. Maximum gold extraction of 74. 16% is obtained under the optimized experiment conditions (*of*): 0. 01mol/L $CuSO_4$, 0. 5mol/L $NH_3 \cdot H_2O$, 0. 1mol/L $Na_2S_2O_3$, initial pH 11. 0 and leaching (*time of*) for 5. 0h at 50℃. Thiosulfate consumption increases with increasing $Cu^+$ and $S_2O_3^-$ concentration, leaching time and (*leaching*) temperature but decreases with an increase of ammonia concentration and initial pH. The effect of pretreating the ore by microwave roasting on subsequent gold extraction with thiosulfate solutions has been examined. When microwave roasting temperature of 500℃ and roasting time of 30 minutes (*was*) are (*used*) appliied, gold extraction from the calcine with copper–ammonia–thiosulfate solution (*reached as high as up to*) can reach as high as 90%. (*Chemical*) Aanalysis shows that (*some*) all organic carbon and partial inorganic carbon (*from the dolomite and free carbon*) in the ore are eliminated (*from carbonaceous gold concentrate in the*) after microwave roasting(*process.*), which may weaken the preg–robbing effect. (*It is believed that pre–robbing effect is weaken, making an increasing gold extraction. Analysis of the calcined concentrate shows t*) The formation of cracks on the surface of calcine particles is observed (*surface*) which may facilitate the permeability of lixiviant during subsequent leaching.

**Keywords:** (*gold*); copper – ammonia – thiosulfate (*leaching*); carbonaceous gold ore; microwave roasting

## 1 Introduction

(原文中此处修改内容较少，且类似问题在其他篇幅出现过，故在此略去部分篇幅)

For carbonaceous gold ore or concentrate, gold extraction by direct cyanidation is usually low due to two types of difficulties. On the one hand, carbonaceous matters can lock up a portion of gold in ore and inhibit leaching. On the other hand, carbonaceous matter also can absorb a large portion of gold–cyanide complex $Au(CN)_2^-$ (namely preg–robbing effect) , (*and cause causing*) the loss of gold in leaching solution. Thiosulfate leaching is considered as a promising way for treating carbonaceous gold ore because of (*a very*) the low affinity of active carbon to gold – thiosulfate complex. Many reports on using thiosulfate solution to extract gold from carbonaceous ores have been published. The patent of applying pressure oxidation and thiosulfate leaching to extraction gold from refractory carbonaceous gold ores was published in 1966. The Newmont Mining company has developed the process of bio–oxidation and thiosulfate heap leaching for low–grade carbonaceous– sulfide gold ores. (*Bin*) Xu et al. (*adopted*) developed an environmentally friendly process of alkaline pressure oxidation–thiosulfate leaching to extract gold from a high C, As and Sb bearing sulfide gold ore. Under optimum experimental conditions, (*The best extraction rate of Au*) gold extraction (*could*) can reach as high as 86. 1%(*. From the above studies, carbonaceous gold ore containing dolomite has not received much research attention.*)

(*The common p*) Pretreatment methods include chemical oxidation, bioleaching and roasting for refractory carbonaceous gold ores have also been studied. Compared with the conventional roasting method, microwave roasting is attractive because of its high efficiency, selected heating, good heating uniformity and reduced environment risk for refractory gold ores. A combined process of microwave roasting and cyanidation was used to extract gold from carbonaceous gold ore containing dolomite, and energy consumption was lower than conventional roasting. (*But it is still unclear what results copper-ammonia-thiosulfate solution extracts gold after microwave roasting treatment.* )

In this research, the process of using thiosulfate solutions to extract gold from carbonaceous ore has been investigated. T(*t*) he influence of copper concentration, ammonia concentration, thiosulfate concentration, initial pH and leaching temperature on gold extraction (*from carbonaceous gold ore containing dolomite were implemented and the consumption of thiosulfate was measured*) and thisosufate consumption was examined. The effect of pretreatment with microwave roasting on subsequent gold extraction (*also*) has also been investigated. Based on these studies, the mechanism and feasibility of leaching carbonaceous gold ore with thiosulfate solutions has been discussed. (*results obtained are valuable for thiosulfate leaching gold from this kind of gold ore or concentrate.* )

## 2 Experiment

### 2.1 Materials

A carbonaceous gold concentrate containing high content of dolomite (*are obtained from*) was provided by Shandong (*gold*) Gold (*group*) Co. of China. (*with*) The concentrate has a particle size of 89.5(0) % (*passing* ) −38μm. (*and then*) The concentrate sample was first dried at 60℃ in air for 12 hours before proceeding to any analysis and experiment. (*Its*) The main chemical composition of the concentrate (*s*) (*are*) is shown (*indicated*) in Table 1. The XRD (X-ray diffraction) spectrogram of (*gold*) the concentrate is shown in Fig. 1. It shows that the carbonaceous gold concentrate is mainly consisted of $SiO_2$, $CaCO_3$ and $CaMg(CO_3)_2$. The concentrate contains 6.6% total carbon with about 0.2% organic carbon. Previous study shows that only 35% gold extraction can be obtained under optimum cyanide leaching conditions, indicating the concentrate is a typical refractory carbonaceous gold ore(*A cyanide leaching experiment for carbonaceous gold concentrate are carried out under the experiment conditions of 25℃, sodium cyanide 0.15% and liquid-solid ratio 3:1. The 64.92% gold extraction is obtained after 36 h. It shows that this gold ore is refractory because gold extraction is less than 80% by conventional cyanide leaching process*).

Distilled water is used in all experiments. The chemical reagents used for experiment, (*such as*) including sodium thiosulfate, ammonia solution (25%), copper sulfate, sodium hydroxide and sulfuric acid (98%) are all analytical reagent (AR) from Sinopharm Chemical Reagent Company (SCRC), (图、表略)

### 2.2 Leaching experiment

Experiments for leaching gold concentrate with thiosulfate solutions (*leaching gold concentrate*) are performed in a 500mL glass reactor using a magnetic stirrer. The agitating rate of the stirrer (*agitation is kept*) maintains at a constant value (300rpm). The (*lixivant*) leaching solution is accurately prepared (*in laboratory*) by mixing (*with*) sodium thiosulfate, ammonia solution, copper sulfate and distilled water. For each test, a portion of 200mL (*of lixivants*) leaching solution is (*added to*) used to leach 50g (*of gold ore*) solid sample. The pulp is maintained at a constant temperature (±0. 5℃) by heating in water bath. The initial pH (*value*) of leaching system is adjusted by addition of 1. 0M sodium hydroxide and dilute sulfur acid. When leaching is completed, the pulp is filtered (*to leaching solution and residue*) by a vacuum filtration machine. In order to calculate the recovery of gold, both leachate and residue (*they*) are immediately subjected to analysis element gold by (*flame*) a atomic absorption spectrophotometry (AAS). (*The cyanide leaching experiment use the same glass reactor, magnetic stirrer and water bath heating as thiosulfate leaching.*)

### 2. 3 Microwave roasting

Microwave roasting experiments are performed in a 100mL ceramic crucible using a high-temperature microwave reactor from (*KMUST*) Kunming University of Science and Technology. A tungsten rhenium thermocouple is used to measure the reactor temperature during the roasting process. All experiments are performed at the same microwave power (2. 50kW) and heating rate. (*These parameters are automatically controlled by reactor. At the end of the roasting process, the samples are immediately took out from reactor.*)

### 2. (3) 4 Analysis

The pH (*value*) of pulp is measured by a pH meter (PB-10, Sartorius). The (*X-ray diffraction*) XRD analyses and the thermogravimetric analysis (TGA) are performed both on (*gold*) the concentrate sample and (*also*) the calcine after microwave roasting (*for concentrate sample*) by X-ray diffractometer (X'Pert Pro, PANalytical B. V.) and thermal gravimetric analyzer (SDT Q600). In addition, The (*Scanning*) scanning electron microscope (SEM, Hitachi, SU8000) is used to observe morphology of solid sample (*particle*). Also the microwave absorption characteristics of carbonaceous gold concentrate are determined by microwave reactor. (*The concentration of g*) Gold content in solution is measured by (*flame atomic absorption spectrophotometry*) AAS (Z-2300, Hitachi). The (*Thiosulfate*) thiosulfate (*concentration*) content in leaching solution is measured by the iodine titration method.

## 3 Results and discussion

### 3. 1 (*Thiosulfate leaching kinetics experiment*) Preliminary leaching tests

T(*he experiment*) ests for leaching the concentrate under different leaching time are firstly carried out(, ) and the results are shown in Fig. 2. Under this experimental conditions, the gold extraction increases with increasing time initially. Gold extraction increases from 44. 8% to 69. 6% when leaching varies from 0. 5 hour to one hour. After leaching for one hour, the gold extraction only increases slightly with increasing leaching time (71. 3% for three hours and 74. 2% for five hours) and then tends to remain stable with an increase of leaching time in the range of five hours to nine hours. Compared to cyanidation process, gold extraction from this carbonaceous ore with thiosulfate system is much higher within same leaching time (e. g., in two hours, 70% for thiosulfate leaching vs 35% for cyanide leaching), indicating the thiosulfate leaching system is more suitable for treating this gold ore. (*However, t*) The thiosulfate consumption gradually increases with increasing time (from 2. 1% for 0. 5 hour(*, 3. 0% for one hour, 7. 3% for three hours, 16. 6% for five hours, 33. 0% for seven hours and*) to 44. 6% for nine hours(*, respectively*)). By comprehensively considering the gold extraction and the thiosulfate consumption, leaching time of five hours are adopt in subsequent tests. (图略)

### 3. 2 Effect of lixiviant concentration

#### 3. 2. 1 Effect of cupric ion concentration

The effect of cupric ion concentration on gold extraction is shown in Fig. 3. It shows that the gold extraction increases from 40. 0% to 71. 9% with an increase of cupric ion concentration when it varies from 0. 0025mol/L to 0. 01mol/L, and then remains stable in the range of 0. 01mol/L to 0. 04mol/L. The thiosulfate consumption gradually increases with increasing cupric ion concentration (8. 4% for 0. 0025mol/L $Cu^+$, 9. 5% for 0. 005mol/L $Cu^+$, 11. 2% for 0. 01mol/L $Cu^+$, 13. 7% for 0. 02mol/L $Cu^+$ and 16. 7% for 0. 04mol/L $Cu^+$, respectively). This result is (*relatively*) consisted with the experimental result by other researchers.

According to literature, (*In the presence of ammonia and copper,*) the dissolution rate of gold is increased in the thiosulfate solution in the presence of ammonia and copper. The main reaction during leaching with copper-ammonia-thiosulfate sulutions can be expressed as following: (公式略)

Based on reaction (1), (2) and (3), the formation of cupric-tetraammine complex in thiosulfate leaching system could accelerate the oxidation of gold. T(*hese reactions show that t*) he dissolution of gold would proceed to the right with an increase of cupric ion concentration. But excessive increase in cupric ion concentration (*could*) would reduce the stability of $Cu(NH_3)_4^+$ and enhance the stability of solid copper compounds such as CuO, $Cu_2O$, CuS and $Cu_2S$, which could cause the loss of copper(Ⅱ) in leaching solution. Moreover, the high cupric ion concentration could also facilitate the oxidation decomposition of thiosulfate due to the undesirable reaction between copper(Ⅱ) and thiosulfate, as shown in reaction (4) and (5).This indicates that the excessive cupric ion concentration is detrimental for gold extraction in thiosulfate solution. (公式略)

(*By comprehensively considering the gold extraction and the thiosulfate consumption, cupric ion concentration of* 0. 01*mol/L is adopt as the optimum cupric ion concentration. This result is relatively consist with the experimental result by other researchers.*) (图略)

#### 3. 2. 2 Effect of ammonia concentration

The effect of ammonia concentration on gold extraction has been examined in the range of 0. 25mol/L to 2mol/L (shown in Fig. 4). It was found that gold extraction increases sharply from 33. 5% to 69. 1% with an increase of ammonia concentration in the range of 0. 25mol/L to

0. 5mol/L and then (*tends to*) increases slightly(*gradually*) from 69. 1% to 71. 9% when it varies from 0. 5mol/L to1mol/L. However, gold extraction slightly decrease when (*continuing increasing in*) in the ammonia concentration further increases ( 1mol/L to 2mol/L) . Moreover, the thiosulfate consumption decreases with an increase of ammonia concentration ( 27. 0% for 0. 25mol/L ammonia, (*22. 0% for 0. 5mol/L ammonia,* ) 11. 2% for 1mol/L ammonia, (*10. 4% for 1. 5mol/L ammonia*) and 9. 3% for 2mol/L ammonia, respectively) .

In the thiosulfate leaching system, ammonia can prevent gold passivation by preferential absorption on gold surface, which improve the leaching kinetics. Ammonia also (*have*) play a (*catalytic*) role in the catalytic dissolution of gold in copper-thiosulfate solution. Moreover, ammonia play a stabilized role in copper( Ⅱ ) by the formation of cupric-tetraammine complex ( $Cu(NH_3)_4^+$ ). Therefore, these could be the reason why gold extraction increased with an increase of ammonia concentration ( 0. 25mol/L to 1mol/L) .

A slight decrease in gold extraction could be because the mixed solution potential and $Cu^+/Cu^+$ reduction potential are more negative at high ammonia concentration than low ammonia concentration. However, the stability of thiosulfate can be enhanced under the high ammonia concentration, resulting in the low thiosulfate consumption ( shown in Fig. 4) . (*By comprehensively considering the gold extraction, thiosulfate consumption and ammonia toxicity, ammonia concentration of 0. 5mol/L is adopt as the optimum concentration for next tests.* ) ( 图略)

### 3. 2. 3 Effect of thiosulfate concentration

The effect of thiosulfate concentration on gold extraction in the copper-ammonia-thiosulfate leaching system is shown in Fig. 5. Initially, the gold extraction increases from 55. 4% to 74. 8% with an increase of thiosulfate concentration ( 0. 05mol/L to 0. 1mol/L) . But then the gold extraction gradually decreases from 74. 8% to 57. 1% with an increas(*e of*) ing thiosulfate concentration in the range of 0. 1mol/L to 0. 4mol/L. (*However, t*) The thiosulfate consumption also increases with an increase of thiosulfate concentration ( 14. 0% for 0. 05mol/L thiosulfate, 16. 6. 0% for 0. 1mol/L thiosulfate, 18. 6% for 0. 2mol/L thiosulfate, 22. 0% for 0. 3mol/L thiosulfate and 22. 8% for 0. 4mol/L thiosulfate, respectively) .

From the point of previous paper(, ) It was reported that an increase in thiosulfate concentration can enhance both the anodic and cathodic half reactions, as shown in reaction ( 5) and ( 6) , . Therefore, gold extraction increases with an increase of thiosulfate concentration (*from 0. 05mol/L to 0. 1mol/L.* ) ( 公式略)

During the thiosulfate leaching of gold, an increase in thiosulfate concentration will accelerate the undesirable reaction between cupric ion and thiosulfate, forming polythionate ( tetrathionate and trithionate) and a series of compounds containing sulfur, which (*cause*) result in the excessive consumption of thiosulfate and the lack of cupric ammonia complex in leaching solution. This is consist with the experimental phenomenon that the color of cupric ammonia complex gradually becomes lighter in leaching solution with an increase of thiosulfate concentration. The decomposition products ( surface sulfur species) of thiosulfate can cover gold surface, which form a passivation film on gold surface, thus resulting in low gold extraction under high thiosulfate concentration.

(*By comprehensively considering the gold extraction and thiosulfate consumption, thiosulfate concentration of* 0.1*mol/L is adopt as the optimum ammonia concentration for next tests.*)(图略)

### 3.3 Effect of initial pH

The effect of initial pH on gold extraction is examined in the range of 9 to 13 which thiosulfate leaching process is usually performed. The test results are shown in Fig. 6. The gold extraction increases from 56.2% to 74.2% with an increase of pH when it various from 9.0 to 11.0, and then (*tends to*) decreases from 74.2% to 70.9% in the range of 11.0 to 13.0. However, the thiosulfate consumption decreases with an increase of initial pH (41.1% for 9.0, (38.3% *for* 10.0), 16.6% for 11.0, (12.2% *for* 12.0) and 5.9% for 13.0, respectively).

In the aqueous phase, the equilibrium speciation of copper in ammonia thiosulfate solution can be represented by reaction (7) – (17). Where $K_1$–$K_{11}$ are the equilibrium constants (here they are selected as $10^{4.31}$, $10^{7.98}$, $10^{11.02}$, $10^{13.32}$, $10^{10.27}$, $10^{12.22}$, $10^{13.84}$, $10^{7.0}$, $10^{13.68}$, $10^{17.0}$ and $10^{18.5}$, respectively at 25℃) (*from Lange's handbook of chemistry.*) The distribution of copper complex in leaching solution at different pH is calculated through program based on mass and charge balances, and the result is shown in Fig. 7. (公式略)

It was observed that the stability region of cupric–tetraammine complex ($Cu(NH_3)_4^+$) is in pH range of 9.0 to 13.0 (*for this leaching solution*). The distribution of cupric–tetraammine complex increases with an increase of pH (9.0 to 11.4). However, a further increase in pH beyond (*about*) 11.4 will decrease the (*distribution*) formation of cupric–tetraammine complex. Some research claimed that cupric–tetraammine complex could accelerate the oxidation of gold, which is of vital importance to increase gold extraction in thiosulfate solution. The cupric–tetraammine complex presents blue color in previous literature. The result is in accordance with the experimental phenomenon that the color of leaching solution changes from shallow blue to deep blue with increasing pH (9.0–11.0) and gradually becomes light with an increase of pH (11.0–13.0).

(*Some research claimed that cupric–tetraammine complex could accelerate the oxidation of gold, which is of vital importance to increase gold extraction in thiosulfate solution.*) Therefore, the optimum gold extraction can be obtained in the region of around 11.0 due to the presence of large amounts of cupric–tetraammine complex.

(*By comprehensively considering the gold extraction, thiosulfate consumption and the distribution of* $Cu(NH_3)_4^+$, 11.0 *is regarded as the optimum initial pH for leaching solution.*)(图略)

### 3.4 Effect of (*leaching*) temperature

The (*experiments for temperature*) effect temperature on gold extraction (*are carried out*) with copper–ammonia–thiosulfate solution(. *The results are*) is shown in Fig. 8. The gold extraction slightly increases from 71.7% to 74.2% with an increase of temperature (30℃ to 70℃), and then decreases from 74.2% to 67.4% with increasing temperature in the range of 50℃ to 70℃. The thiosulfate consumption increases with (*an increase*) increasing of temperature—under experimental conditions(*(5.8% for 30℃, 10.2% for 40℃, 16.6% for 50℃, 32.2% for 60℃ and 41.2% for 70℃, respectively)*).

The increase (*in the leaching*) of temperature (*could be attributed to increase gold extraction by*) may facilitate (*enhancing*) gold (*oxidation half reaction,*) dissolution through preventing passivation (*phenomenon*) on particle surface and reducing the mixed potential of leaching gold. (*But*) However, (*excessively high*) increasing temperature (*could sharply*) would also increase thiosulfate decomposition by (*facilitating*) promoting the reaction between cupric ion and thiosulfate, resulting in the loss of cupric ion and thiosulfate in leaching solution. It should be noted that a portion of decomposition products (namely surface sulfur species) could form a passive layer on the gold surface, thus inhibiting the contact (*with*) of gold (*and*) with leaching solution. (*These may be the reason why gold extraction is reduced at high temperature.*) Thus, under experimental conditions, a decrease in gold extraction was observed when the leaching temperature exceeds 60 C.

(*Owing to the volatility of ammonia at high temperature, 50℃ is selected as optimum leaching temperature.*)(图略)

### 3.5 Effect of pretreatment with microwave roasting

In order to examine the potential of (*improve*) improving (*the*) gold extraction (*of*) from this carbonaceous gold (*concentrate*) ore, a series of tests using microwave roasting (*experiments*) to pretreat the ore before thiosulfate leaching are performed (*before thiosulfate leaching*). (*The operating conditions of microwave roasting chooses low roasting temperature (300℃, 400℃ and 500℃) and short roasting time (10 minutes and 30 minutes).*) The heating rate curves of the concentrate is shown in Fig. 9. After the microwave roasting pretreatment, calcine are leached under the optimized thiosulfate leaching conditions: 0.01mol/L $Cu^+$, 0.5mol/L $NH_3 \cdot H_2O$, 0.1mol/L $S_2O_3^-$, initial pH 11.0 and leaching time for 5.0 hours at 50℃. (图略)

The test results (*of thiosulfate leaching*) are shown in Table. 2. Compared to (*74.16% gold extraction of*) direct thiosulfate leaching, (*it is no obvious change in*) gold extraction exhibits virtually no change when the ore was pretreated with (*of*) microwave roasting (*sample*) under 300-400℃ for 10 minutes (*under 300℃ roasting temperature*). (*However, gold*) Under same experimental conditions, gold extraction (*gradually*) increases from 74.9% to 82.3% (*with an increase of*) when the roasting temperature (*from*) increases from (300) 400 to 500℃(, ). and an increase in gold extraction also increases with roasting time from 10 minutes to When the roasting time of 30 minutes is used(.), (*When microwave roasting temperature of 500℃ and roasting time of 30 min is used,*) gold extraction (*can reach as high as 90%*) increases from 87.4% for 300 C to 90.2% for 500 C.

The XRD pattern and TGA curve of (*carbonaceous gold*) the concentrate and (*its*) the calcine from microwave roasting (*sample*) are (*respectively*) shown in Fig. 10 and 11, respectively. It (*is observed*) shows that the phase composition of gold concentrate has no change in microwave roasting process.

(原文中此处修改内容较少，且类似问题在其他篇幅出现过，故在此略去部分篇幅)

## 4 Conclusions

(*The l*) Leaching of a carbonaceous gold concentrate containing high content of dolomite with copper–ammonia–thiosulfate solutions has been investigated. (*and results show that the*) Under optimum(*al*) experimental conditions ($CuSO_4$ 0.01mol/L, $NH_3 \cdot H_2O$ 0.5mol/L, $Na_2S_2O_3$ 0.1mol/L, initial pH 11.0 and leaching time for 5.0 hours at 50℃), gold extraction with thiosulfate solutions can reach 74.16% which is much higher than with traditional cyanidation process. The optimized thiosulfate leaching conditions for this gold concentrate is (*$CuSO_4$ 0.01mol/L, $NH_3 \cdot H_2O$ 0.5mol/L, $Na_2S_2O_3$ 0.1mol/L, initial pH 11.0 and leaching time for 5.0 hours at 50℃.*)

(*During the thiosulfate leaching o*) Uner experimental conditions, (*f carbonaceous gold concentrate*), thiosulfate consumption increases with an increase of $Cu^+$ ((*0.0025–0.04mol/L*)) and $S_2O_3^-$((*0.05–0.40mol/L*)) concentration, leaching time ((*0.5–9.0 h*)) and leaching temperature ((*30–70℃*)) but decreases with an increase of ammonia concentration ((*0.25–2.00mol/L*)) and initial pH ((*9.0–13.0*)).

Microwave roasting pretreatment (*is*) proves to be an effective (*method*) way to improve (*gold extraction for*) leaching gold from (*of*) carbonaceous (*gold*) ore with copper–ammonia–thiosulfate solutions. When microwave roasting temperature of 500℃ and roasting time of 30 minutes is applied, gold extraction can reach as high as 90%. By comparing the XRD pattern and TGA curve of carbonaceous gold ore and its microwave roasting sample under the roasting temperature of 500℃, it is observed that the phase composition of gold concentrate has no change in microwave roasting process. The chemical analysis of carbon element shows that all organic carbon and particle inorganic carbon in the ore are eliminated after microwave roasting, which may weaken the pre–robbing effect. The SEM photograph of the calcined concentrate shows the formation of cracks on the surface of calcine particles which may facilitate the permeability of lixiviant during subsequent leaching, thereby increasing gold extraction.

(*From an industrial point of view, t*) The experiment results show that the process of microwave roasting pretreatment – thiosulfate leaching has (*promising industrial prospect*) great potential for treating carbonaceous gold concentrate. (*The future work will be focused on improving gold recovery, decreasing thiosulfate consumption and reducing microwave power.*)

**References:**

(参考文献略)

# 参考文献

[1] YANG J T. 科技英语写作［M］. 王多，译．上海：复旦大学出版社 2012.

[2] 泰狄辉．实用英语写作［M］. 上海：上海外语教育出版社，2001：87.

[3] 泰狄辉．科技英语写作高级教程［M］. 第二版．西安：西安电子科技大学出版社，2011.

[4] 吴小力．冶金科技英语口译教程［M］. 北京：冶金工业出版社，2013：225.

[5] 泰狄辉．科技英语写作［M］. 北京：外语教学与研究出版社，2007.

[6] 泰狄辉．科技英语写作教程［M］. 西安：西安电子科技大学出版社，2001.

[7] 赖世雄．赖世雄经典语法［M］. 北京：外文出版社，2009：273.

[8] XIE F, ZHANG T A, DREISINGER D, et al. A critical review on solvent extraction of rare earths from aqueous solutions[J]. Minerals Engineering, 56: 10~28.

[9] XIE F, DONG K, WANG W, et al. Leaching of mercury from contaminated solid waste: a mini-review[J]. Mineral Processing and Extractive Metallurgy Review, 2020, 41(3): 187~197.

[10] SWALES J M, FEAK C B, et al. Academic writing for graduate students [M]. Detroit: The University of Michigan Press, 1994.

[11] GLASMAN-DEAL H. Science Research Writing: For Non-Native Speakers of English [M]. London: Imperial College Press, 2010.

[12] MORLEY J. Academic Phrasebank[EB/OL]. www. phrasebank. manchester. ac. uk.